GOVERNING THE WIND ENERGY COMMONS

VOLUME FIVE

RURAL SOCIOLOGICAL SOCIETY

Rural Studies Series
Sponsored by the Rural Sociological Society
Edited by Spencer D. Wood, Kansas State University

GOVERNING THE WIND ENERGY COMMONS

Renewable Energy
and Community Development

Keith A. Taylor

WEST VIRGINIA UNIVERSITY PRESS | MORGANTOWN 2019

Copyright © 2019 by West Virginia University Press
All rights reserved
First edition published 2019 by West Virginia University Press
Printed in the United States of America

ISBN
Cloth 978-1-946684-84-4
Paper 978-1-946684-85-1
Ebook 978-1-946684-86-8

Library of Congress Cataloging-in-Publication Data
is available from the Library of Congress

Cover design by Than Saffel / WVU Press

To Rebecca, for your unending tenacity, love, and enduring support for changing the world. To my mom and grandma Betty, for shaping me to be the person I am today. To my closest friends, Josh Bright, Chris Lempa, and Jacqueline Hannah. And to my heroes, all of those who strive simply to have a meaningful, fulfilling life.

CONTENTS

ACKNOWLEDGMENTS

The book you have in your hands could not have become a reality were in not for a wonderful supporting cast of great characters working with me and supporting my work.

I would like to acknowledge first and foremost the editorial staff with West Virginia University Press's Rural Studies Series. The team has remained incredibly helpful and patient throughout the entire process. I have learned a great deal through this process through the communiques with the press team. I cannot thank you all enough for your patience.

I would like to express my gratitude to the Rural Sociological Society, which recognized my work as a meaningful and worthy contribution, not to mention the many wonderful peer reviewers who provided input throughout the process.

Additionally, I would like to thank my team of advisers at the University of Illinois and the University of Kentucky, Stephen Gasteyer, William Sullivan, Gale Summerfield, and Patrick Mooney, who shepherded through an earlier version of this work.

Rural America is wind energy country. Vast swaths of the rural countryside reside at the intersection of plentiful wind and land, affordable electric grid access, and a jobs-hungry workforce. These features are essential for building the next generation of clean wind-energy infrastructure, and rural communities are at the forefront of the wind energy frontier (Brown, Nderitu, Preckel, Gotham, & Allen, 2011, p. 173).[1]

There are hundreds of wind farms across resource-rich terrains of the United States, many having operated for decades. These wind farms require enormous infusions of capital and collective action from national, state, and local communities, leading many to conclude that wind energy development confers direct benefits to rural communities. This cuts against the dominant narrative that sees a rural America in despair, with few resources, left to hollow out and evacuate for more affluent urban regions.

Collectively, billions of dollars have been sunk into wind energy development across rural American communities. With that level of investment, it is easy to see how one would presume that wind energy development could be harnessed for the collective benefit of wind energy communities. Wind energy is, after all, quite a lucrative enterprise, with profits to spare. What, then, do the years of experience tell us about the community benefits to rural livelihoods and civic institutions? Who is developing these wind resources, and do they share in the communities' concerns? What is more, is this boom in wind energy development a disruption to the status quo, or more of the same? The big question is, how does wind energy development influence community economic development?

The answers to these questions are of enormous importance for understanding how to improve the livelihoods of rural dwellers, first, by understanding how national-level policies could be used as a "two-fer" to meet the objective of converting from carbon to renewable energy, harnessing policy for the well-being of those frontline communities tasked with shepherding the policy to success; and second, by understanding how wind energy policy's reliance on public-private partnerships is engaging the appropriate array of partners.

Overview

The Growing Reach of Wind Energy

Wind is the leading source of renewable energy globally (U.S. Energy Information Administration, 2016). Beginning in 2011, wind energy was projected to grow at an annual rate of 39 percent over the next five years (World Wind Energy Association, 2010), with a doubling of total installed wind energy capacity every two and a half years (Madison, 2010). Actual global wind energy growth has been a bit slower than the projected: 20.3 percent in 2011, 16 percent in 2012 (World Wind Energy Association, 2012), and 16.8 percent in 2015 (World Wind Energy Association, 2015). However, wind's total reach into the global energy portfolio mix continues to flourish by leaps and bounds.[2]

Energy governance has received little scholarly attention (Aitken, 2010) despite the "extraordinary importance of energy" in current affairs (Florini & Sovacool, 2009, p. 5240). Research on wind energy development processes is mostly relegated to economic modeling and policy analysis, with little attention paid to the social and community outcomes (see, e.g., Kildegaard & Myers-Kuykindall, 2006; Loomis & Carter, 2011). Readily available data come primarily from environmental, industrial, or oppositional interests, which are firmly in the "for" or "against" camps as opposed to being impartial referees that might provide a more nuanced perspective. This means that community economic developers and government policy makers have a narrow scope of data with which to inform wind energy development decisions. We simply do not have a great deal of evidence on how communities and wind energy interests coexist.

If we are intentional, and if we position wind energy communities to have a seat at the development table, there is an opportunity to recognize real economic and environmental justice for residents of rural America. Opposition to legacy carbon energy (natural gas fracking, coal mining) provides community economic development interests an opening to leverage wind for community-level benefits (McKibben, 2007). Plus, the relative infancy of renewable energy—and its price competitiveness relative to fossil fuels—provides a brief window of opportunity to elevate forms of community ownership of wind energy and to utilize surplus wealth for enhancing citizen self-governing and community economic development capabilities (Sen, 2011). This would have the effect of elevating community voice and power by both diverting a sizable portion of the energy sector's revenue to community-stewarded governance and shutting off revenue streams to the extractive fossil fuel interests involved in undermining community voice.

Despite the assurances of supporters, wind energy's transformative potential has yet to crystallize. A quick scan of wind energy association websites reveals that the multinationals involved in wind energy predicate their value primarily on job creation, growth through profitability of the firm and its spillover into the community, and new sources of tax revenue. Rarely is the empirical evidence weighed against the claims of wind energy development outcomes, nor do the claims account for the perspectives of individuals who have firsthand exposure to wind energy development in their community. Community developers, planners, and energy policy makers may be missing a significant opportunity to enhance community livelihoods and increase policy adoption by taking the claims for granted and ignoring the evidence.

The State of Rural Communities

The enduring economic plight of rural American communities is real. Today's rural economy and this generation's working class are facing a great deal of uncertainty and significant downward pressure on their quality of life because of powerful social forces seemingly beyond individual control. Economic opportunities for rural households are being crowded out by a changing manufacturing sector beset by the encroachment of automation technologies and of ever-centralizing agricultural concerns; jobs are going away, and they are not coming back. While the postwar and baby boom generations brought a progressive era of government expansion into public services, the last three decades have shown that rural communities can no longer count on assistance from the state. Urbanization—the outflow of rural residents to population-dense regions with greater economic opportunities—and the widespread acceptance of small government ideology mean that federal and state governments continue to cut vitally important funding and programming for rural communities. More recently, the Great Recession has advanced austerity, choking off rural redevelopment resources (health care, education, and other public goods; U.S. Council of Economic Advisers, 2016), inflicting pain on the livelihoods of rural dwellers during already challenging times. As if that weren't enough, climate change compounds the problem in other parts of the rural economy, shifting the predictably reliable boundaries where agricultural, forestry, and recreational industries flourish, increasing stress on natural-resource-dependent communities (Melillo, Richmond, & Yohe, 2014); the Trump administration's hostility toward mitigating climate change promises to speed these processes along. This means that long-stable incumbent industries are being exposed to combinations of economic and environmental disruptions. Our systems of governance are failing, and rural communities are at the front line of the systemic collapse.

The totality of environmental disruption and receding economic opportunity confronting many rural American communities contributes to increased urbanization. The loss of human capital, critical for collective action and public entrepreneurship, further limits the capacity of rural communities to recover and thrive. Rural residents are left feeling they have no recourse for countering the forces of change; they are mere subjects, not agents, unable to actively shape their future. An eminent feeling of hardship, crisis, and powerlessness emerges.

Because the global economy runs on a foundation of accessible, inexpensive, abundant fuels, rural communities have often gone down the path of energy development to secure some degree of spillover benefit. Not all of the outcomes have been positive. The community-level impacts of fossil-fuel development are well known, including documented historical patterns of environmental and economic exploitation of rural communities by outside private interests (Gaventa, 1982). The exploitation then spills over into the host communities, highlighting the disruptive potential of absentee-owned energy development on host communities (Albrecht et al., 1982).

The enduring allure of attracting these incumbent industrial interests persists, even as broader public sentiments shift. Resource-rich communities perceive the need to accommodate an influx of externally provided economic investment to increase tax revenue and employment opportunities. The traditional model of economic development—"smokestack chasing"—presents a Faustian bargain for community economic developers: the community receives jobs and tax revenue but in return is handcuffed to a dominant economic actor whose interests do not align with the community's.

Ongoing opposition to coal and the increasing demand for renewable energy is causing a wave of disruption across the energy industry. A number of the most prominent coal producers have filed for bankruptcy, initiating sizable layoffs (Cohan, 2016). What are energy-producing communities to do when an investor-owned firm, the engine of the local economy, shuts its doors? The community, with "no absentee ownership to safeguard" (Veblen, 1994, p. 5), is left powerless. The long-standing Faustian bargain is losing its luster. Can we learn from past experience in carbon energy development in preparation for the renewable energy future that incorporates an empowering community economic development?

The Wind Energy Prospect

Rural communities are at the center of another mass expansion of energy development, this time in the guise of sprawling wind turbines and solar panels. The renewable energy boom is ushering in a transformative era for communities rich in heretofore unexploited resources. Numerous government incentives and subsidies are available, helping make wind a growth market, an area of economic stimulus not readily found in other rural industrial sectors. Fortunately, access to state assistance and community investment opportunities for wind energy projects is not wholly out of reach of rural communities. Should the proper institutional actors coalesce, the outcomes could be markedly positive for energy-producing communities. Wind energy development has the potential to drive a new economic development paradigm in rural communities founded on collective empowerment, citizen-led climate change mitigation, and the conversion of the electric sector to domestically sourced clean, renewable energy. This time, things could be different for rural America.

It is essential to understand the trends in wind energy development, to contextualize how and why the wind energy development experience may influence the day-to-day lives of those residents living in the host communities.

- What do we know of the development claims of wind energy firms? Do the institutions involved in wind energy development benefit their host communities, and do their claims hold up to scrutiny?
- Do the public policy pathways intentionally maximize social welfare and community gains from this new energy boom, or are such outcomes left to market forces and the whims of the owners?

Renewable energy "boomtowns" will look different from the coal and oil boomtowns of old. Instead of a centralized generating facility owned by the energy company, the wind turbines are sprawled over vast expanses of land owned by numerous stakeholders, including public lands and private farmland, necessitating public participation. Once built, the local wind energy infrastructure can be expected to extract energy from wind resources for twenty-five to fifty years, transforming the local landscape and community livelihoods in perpetuity. Unlike carbon-based generation, wind energy is an inexhaustible source of low-cost energy. The zero-cost aspect of the fuel source better controls for market volatility, conferring more foreseeable utility rates for consumers and reliable revenue streams for wind energy owners. Communities hosting wind farms could harness a level of stable, long-term planning capacity not found in communities reliant on fickle coal, gas, and oil markets.

It is critical to know the actors involved in wind energy development and understand their underlying motivations that shape community economic development outcomes. Wind-energy development interests are differentiating themselves from carbon-emitting energy producers by linking wind energy to social and environmental causes, promising a social benefit. Public policy is silent on the issue of general social welfare, institutional diversity (public vs. private), and ownership. Moreover, the pledges made by wind energy companies are abstract and opaque at best, which is problematic considering many of the companies involved in wind energy are also engaged in highly controversial fracking and coal mining endeavors — why should one presume their underlying motivations to be any more socially responsible for wind energy? Considering that wind energy is an essential part of the U.S. energy strategy, mutuality and trust between the wind energy developers and local community should be factored in toward the continued advancement of wind energy adoption. Should we disengage with local communities, we risk revolts across rural American communities. Why couldn't public policy explicitly account for participatory community governance and economic benefit? If government policy is incentivizing wind energy development, why can't policy explicitly incentivize community ownership in wind energy?

The answer is, it can, and it does.

Ownership and Public Policy:
Investor-Owned Utilities and Electric Cooperatives

The thesis of my argument is that the options available for wind energy ownership will result in differing community economic development outcomes. Further, this study contends that renewable energy policy explicitly prioritizes profit-seeking, investor-owned utilities (IOUs), entrusting investors and their agents in big business to provide for the public interest.[3] This a missed opportunity to (a) engage the general public in renewable energy entrepreneurship and (b) experiment with modes of public-private partnerships that not only enhance the primary policy objectives (climate change mitigation and energy security) but also offer enhanced spillovers in community economic development. This study is concerned with the differentiating features of IOUs and cooperative wind energy firms.

IOUs are firms collectively owned by investor-shareholders. Investor-owned firms are attractive for policy makers because they allow for rapid collective action. A wind energy firm directs a diversity of interests to pool their resources toward a laser-focused project: the development of a wind farm. The shareholders can be local to global in scope, attracting an abundance of investment capital needed to grow wind energy assets. While shareholders have formal control of the firm, management effectively wields power to maximize a return on investment. The separation of ownership and control—absentee ownership—and the for-profit orientation of the IOU inform the strategic direction of the management team toward profit maximization (Hansmann, 2000). Management is primarily concerned with meeting the dictates of the shareholders, as opposed to the interests of the firm's other stakeholders, such as workers, consumers, and the surrounding community. That said, this leads to economic efficiency, expert leaders, and rapid policy adoption, all of which stands to benefit climate change mitigation.

Community ownership stands in contrast to the IOU model. One community-based ownership model of wind energy that has received little attention in the

Table I.1. International Co-operative Alliance cooperative principles and values

Principle	Values
CP 1: Voluntary and open membership[a]	Self-help
CP 2: Democratic member control	Self-responsibility
CP 3: Member economic participation	Democracy
CP 4: Autonomy and independence	Equality
CP 5: Education, training, and information	Solidarity
CP 6: Cooperation among cooperatives	Honesty
CP 7: Concern for community	Openness
	Social responsibility
	Caring for others

[a]Membership is ownership in a cooperative, translating to greater voice in governance and rights to residuals.

scholarly literature is the cooperative. That a body of evidence around understanding electric cooperatives does not exist is vexing. Cooperative-owned businesses, in general, have been linked to enhanced development outcomes (Deller, Hoyt, Hueth, & Sundaram-Stukel, 2009; Fazzi, 2011; Restakis, 2010), particularly in regions with robust cooperative economic sectors (Putnam, Leonardi, & Nanetti, 1993). Proponents of cooperatives point to the design of the co-op, stemming from the International Co-operative Alliance (ICA) cooperative principles and values (CPs; see Table I.1), as the source of the cooperative's community-oriented behavior.

Currently, almost all of the rapidly expanding wind-energy industry assets are investor owned, with only two utility-scale cooperative wind farms existing in the entire United States as of 2013.[4] Consider that 3,150 utilities comprise the U.S. electric sector (USDA Rural Development, 2000). According to the National Cooperative Business Association, 905 of the 3,150 (29 percent) are electric cooperatives, 220 are electric IOUs, and the remainder are municipal and other forms of ownership (NRECA International, 2016, p. 32). Despite the seemingly small share of ownership relative to other institutional options, cooperatives own and operate over half of all electrical distribution lines (which must cover greater spans in rural communities than in dense urban areas). The electric cooperatives encompass 70 percent of the American landmass, delivering 11 percent of the nation's electricity to 42 million people in forty-seven states, employing a workforce of 70,000 (see Table 4.2, in Chapter Four, for statistics comparing the two sectors).[5] It is striking that electric co-ops have a virtual zero market share of wind energy; perhaps policy is impinging ownership diversity. Policy makers tend "to take it for granted that, in the absence of government intervention, large-scale enterprise will be organized in the form of investor-owned firms" (Hansmann, 2000, p. 1) and that investor-ownership would be ideal for economic development. However, merely presuming positive community economic development outcomes result from the privileging of IOUs within the national-level energy and economic development policy does not make it so.

Where Do We Go from Here?

Neither model of wind energy ownership, cooperative or IOU, has been critically analyzed for its community economic development effects. "In fact, because collectively-owned firms are so numerous and so varied, and because they are subject to the forces of market selection, for many purposes they provide a better means of studying political institutions than do than do the governmental entities that are the usual focus of work in political science" (Hansmann, 2000, p. 4). The research presented in this book adds to the energy, community economic development, public policy, organizational, and new institutional literature through an analysis of the community economic development outcomes of wind energy development. This research intends to fill that knowledge gap and to increase understanding of the implications of this rapid rate of growth on wind-energy host communities, as well as the ramifications of the dominant use of one type of institutional ownership over another (institutional monocropping; Evans, 2004).

This study comprises case studies of two types wind energy firms and their host communities: PrairieWinds in Ward County, North Dakota, a cooperatively owned wind farm operated by the Basin Electric Power Cooperative; and Twin Groves wind farm in McLean County, Illinois, operated by the investor-owned Horizon Wind Energy. The two cases were chosen based on data culled at the outset of the research in 2010. Illinois was of interest because the state was nationally ranked sixteenth in total wind energy resources but sixth in actual installed wind-energy generation capacity (AWEA, 2012b). North Dakota, on the other hand, was ranked number one nationally concerning total wind energy resources but tenth in actual installed wind-energy generation capacity (AWEA, 2012b).

The North Dakota case of Ward County is of particular interest because it is host to the first fully cooperatively owned and operated utility-scale wind farm, one of only two in the nation. The development of PrairieWinds involved the collective action of multiple cooperative organizations, as opposed to a single investor-owned entity taking the lead. What factors drove North Dakota co-ops to be their sector's leaders in deploying wind energy assets? The community of McLean County, Illinois, by contrast, hosts an investor-owned wind farm in a state with only IOU wind farms. While the firms both operate within the same body of federal laws and regulations, the case of McLean County in Illinois serves as a comparison to assess how the social norms, markets, and local rules vary and influence the outcomes of wind energy development.

Comparing the differences in behavior between the two types of ownership—investor-owned versus cooperative—not only is crucial for organizational scholars of each type of firm but also allows us to better assess the applicability of both ownership models within given contexts. Do such firms operate in a manner that significantly alters the delivery of public policy objectives, managerial efficiency, and market competitiveness? Such comparison provides a better understanding of how the peculiarities of history lent themselves to institutional evolution and the manner in which laws, statutes, and regulations may advantage one institutional type above another.

Let's not overlook what is often obscured in public policy: policy's community impact. What happens when we have a world in which for-profit economic interests are explicitly writing the rules of the game, while those organizations purporting to be community oriented—cooperatives—are relegated to the dustbin by policy makers? We cannot miss the forest for the trees, meaning that, in broader economic development policy (the forest), often we think that the firm should be the prime beneficiary of privilege and is the one presumed to be a natural outcome of market processes, the for-profit endeavor (the tree, in our metaphor). Yet we know that forests are themselves ecologically diverse, comprising numerous trees varieties. Also, depending on the composition of trees in the woods, the forest will exhibit differing attributes of resilience. We must not overlook the study of ownership varieties, their politics, and their spillovers into the community, which promises to deliver insight into political institutions, if we are to pretend to have any concern for the community-level impact (Hansmann, 2000). Comparative analyses of ownership types enhance the strength of the findings, whereas the study of a single organizational model provides "too few degrees of freedom and too little variance, to assess the influence of key variables" (p. 3).

This study takes a New Institutional approach—analyzing institutions and how their rules-ordered structures affect social outcomes—to understanding

- how wind energy development interacts with local-level community economic development,
- how the ownership model influences the actions of the wind energy firm, and
- how the diverse systems of local, state, federal and market-based governance influence these interactions (Ostrom, 2005).

Research tools crafted from the foundational work of Bloomington School political economists Elinor Ostrom and Vincent Ostrom, and the international networks of scholars associated with their research institute, are applied to contextualize the interactions among these communities. This contextualization is performed through analysis of the multilayered connections between communities, wind energy organizations, governance institutions (state regulators), individual actors (e.g., developers and landowners), and social systems more generally (Ostrom, 2005, p. 13).

We understand the community influence of wind energy development through the exploration of localized social and governance structures and the actors embedded within them. The two case studies provide in-depth community-level analyses that may inform future efforts in collecting longitudinal data. The emphasis is on how the wind energy utilities interact with the community social structure and how they alter the observed social structure through wind energy development and operations. Being on the ground to analyze these interactions may, in fact, provide insights on phenomena not accounted for, specifically, the tension redirected back at the utility by the community actors in the interaction process—in other words, change may come from the bottom up. It is vitally important to collect firsthand accounts from those who live this experience, as opposed to industry interests alone, to gain a more nuanced perspective of the impact of wind energy on local communities. The two case studies incorporate the perspectives of those who live within the host community to paint a complete picture of the interactive processes and community outcomes (Poteete, Janssen, & Ostrom, 2010, p. 52). The research incorporates perspectives through visits to and observations of the communities, interviews with relevant community members, and archival analyses of local documentation such as newspapers.

The central question, *How does wind energy development influence community economic development?*, requires analyses of many socially ordered relationships. The relationship of interest in this study is the interaction between a wind energy utility and the host community. The study assesses the institution of the wind energy firm. Further impacting the local community and the wind farm are other connected, adjacent processes, such as the market, government regulation, civic participation, and other prevailing trends and policies (McGinnis, 2011b).

Three tiers of questions arise out of the central research question, in order of priority: the firm level, community-level policy, and market level.[6] This book addresses three subquestions to evaluate the implications of the current policy and market orientations and to assist community leaders, development practitioners, planners, and policy makers in making choices better suited to the needs of those most impacted by wind energy development.

Firm Level

How does the ownership model of the wind energy firm affect community economic development? This study begins to fill the knowledge gap on wind energy, ownership diversity and choice, and community economic development by comparing the community interactive aspects of investor-owned and cooperative owned wind energy through a critical comparison of the two cases. The study is designed not to presume that one ownership model has a higher value than the other (e.g., perhaps the electric energy sector is too complicated for fallible human beings to govern democratically?). This orientation provides insights into the community economic outcomes of wind energy development, differentials stemming from ownership, and the manner in which wind energy firms do or do not serve their host communities.

- How does the ownership structure of the firm influence the community outcomes?
- Does the firm's ownership structure impact the capacity of individuals to act collectively to meet their common ends (Ostrom, 2005), and if so, how?
- How does the firm work with the local elites and marginalized populations?

We know from the environmental sociology literature that the motivations driving for-profit energy interests are mostly financial, not social, with little expressed concern for the well-being of residents (Adamson, 2008; Freudenburg, 1984, 2008).

- Would local ownership result in more socially responsible, community-beneficial outcomes and, by extension, higher rates of community acceptance of wind energy than that observed from IOUs?

If local ownership provides a significant incentive for the development of carbon-neutral wind energy, then this should be of great interest to policy makers seeking a rapid conversion from fossil fuels to renewables; from a national perspective, the more hands on deck, the better.

The questions posed at the organizational and community levels of analysis attempt to aid understanding of community interaction and engagement through the perspective of the firms:

- How does the utility's ownership engage the community? How does it view its responsibility to the community that is hosting the wind energy generation assets, and why?
- How do the internal organizational logics influence the degree of community economic development?

Community Level

How does wind energy development influence and interact with local community social structures? The classical energy boomtown research of the 1970s and 1980s dealt with fossil-fuel energy interests, critically assessing industry claims of enhanced material well-being within host communities. This body of research assessed the claims of job creation, increased local tax revenue, and economic development in disadvantaged rural areas (Bacigalupi & Freudenberg, 1983; Black, McKinnish, & Sanders, 2005; Freudenburg, 1979). Further research adds nuance to this critical literature around boomtowns, finding evidence of community benefit (Brown, Dorins, & Krannich, 2005). The disparity in assessments indicates a lack of knowledge and relevant strategies to utilize new knowledge for community adaptation during times of economic disruption and opportunity.

Now, a few decades later, the United States is in another boom cycle of energy development, pushed simultaneously by the need to expand electric generation and transmission capacity while converting to more environmentally sound sources of fuel. The energy industry and its advocates are making similar economic development claims yet again about benefits to host communities (Vilsack & Chu, 2010), despite scant research on the community economic outcomes from wind energy development.

The process of wind energy development exhibits many traits similar to the fossil fuel boom, though with obvious deviations. Much of the advocacy on behalf of wind energy is coming from industry, emphasizing issues of environmental and economic social justice, issues of concern for many advocates of community economic development. Renewable energy writ large has been criticized for greenwashing, serving as a Trojan horse for legacy energy to maintain its franchise and reboot its image (Miraftab, 2004, p. 91; Pellow & Brulle, 2005). Communities have good reason to be skeptical about wind energy development.

Studies on the community impact of wind farms have found the claims of positive community economic development spillover to be overstated (Finzel & Kildegaard, 2013; Loomis & Carter, 2011). The economic boost provided to communities is front-loaded, having the most significant effect during the construction phase. The long-term operations phase contributes to the local tax base and land-lease payments but provides relatively few jobs and may be contributing to capital flight due to the extraction of profits into shareholder's hands. Detailed community-level data are difficult to come by; longitudinal analyses are unavailable, since the industry is in relative infancy. The lack of knowledge is problematic for communities pressured into making long-term development decisions (land-lease contracts for turbines can range from twenty to fifty years) based on company assurances predicated on short-term projections of return on investment.

Policy and Market Level

How do the local to national governance systems of the United States influence local-level wind energy development? Wind energy development is a creature of private-public partnerships, requiring significant capital outlay, a considerable portion being offset by government subsidies. The wind energy industry receives substantial federal and state subsidies in the name of promoting national energy independence from "foreign" energy sources (N.C. Clean Energy Technology Center, 2012). However, wind energy companies interact with the government not only on matters of capital procurement but also on matters of regulation, land management, and industry-dictated connectivity of grid infrastructure. Many states have established aggressive renewable energy portfolio standards (renewable portfolio standards, RPSs), mandating that a specific share of electricity be generated from renewable sources within a given time frame, further stimulating demand for wind energy. We must account for the significant role that government plays.

- How do the policies of the market, state, and other governance institutions influence, privilege, or impede the ownership model of the wind energy firms?
- To what extent can the firm utilize extracommunity mechanisms of power and influence to subvert local policy makers ("level-jumping" behavior; McGinnis, 2011b)?
- What are the unique impediments to development faced by the two ownership models of the wind energy utility (the IOU's for-profit status vs. the cooperative's not-for-profit categorization)?

Book Structure

The research on wind energy has yet to assess the community economic development differentials of one ownership type versus another (the ownership diversity question) and if one model of ownership may be preferable depending on the pre-defined policy outcomes. Of additional interest is the cooperative question, namely, what discourages development of cooperative-owned wind farms? Using the three broad subquestions outlined above, the rest of the book evaluates the implications of the current policy and market orientations, to assist community leaders, development practitioners, planners, and policy makers in choosing options better suited to the needs of those most impacted by wind energy development.

Chapter One reviews the literature within which the broader analysis is couched and lays out the study design. Chapters Two and Three present the two case studies of McLean County, Illinois, and Ward County, North Dakota. Each study covers how wind energy development influences local-level community economic development. Chapter Four compares the two case studies, analyzing the potential importance of ownership on their host communities. Chapter Five then concludes with proposals for future research on renewable energy and cooperative businesses, and mobilization strategies for community organizers interested in leveraging wind energy development for economic and environmental justice.

CHAPTER ONE

Community Development and Institutional Fit

French philosopher Alexis de Tocqueville observed that the mores of the citizenry and the institutions at their disposal (e.g., churches or community groups) have "proven more essential to the continued development of democracy than the details of its legal system or its physical location" (McGinnis, 2011a, p. 27). Religious congregations provide spaces for people to convene, community groups teach people how to work together, and cooperatives provide a venue in which to govern economic and civic spaces collectively.

This book focuses on investor-owned firms and cooperatives operating at the crossroads of individual power and capability, the local communities in which those individuals are situated, and the market economies that provide goods and services to those individuals and their communities. Powerful economic and political interests are connected to these businesses. Renewable energy interests are no different, with Wall Street brokers (e.g., Goldman Sachs) and incumbent energy producers (e.g., Duke Energy) playing a significant role in implementing renewable energy policy.

A manufacturer could bring in hundreds of high-paying jobs, or the siting of a big-box retailer may result in the closure of numerous locally held family businesses. In a similar vein, wind energy interests promise prosperity in the form of hundreds of jobs and reinforce collective responsibility in fending off climate change.

- What should a community expect from these projects, how should community members prepare, and what is the responsibility of business interests in delivering community benefit?
- How do we understand how communities interact with significant economic development projects? Do these communities take an active role in shaping economic development impacts? Are the material benefits broadly distributed throughout the community, do local elites reap most of the gains, or do the benefits accrue to the firm and its ownership?
- How is the capacity of local community members to self-govern and engage in community economic development safeguarded to optimize the spillovers provided from an influx of economic development?

Business development outcomes can be a mixed bag. A thriving community economy requires business development and innovation, particularly with the looming threat of an automated economy promising to eradicate a broad array of formerly stable career pathways. Uncertainty becomes the norm if the community does not intentionally position itself to shepherd economic development outcomes. Perhaps the provision of ownership and governance rights in the project could assure enhanced community economic development outcomes. Moreover, perhaps community ownership through the cooperative business model removes a great deal of community economic development uncertainty.

This chapter establishes the theoretical lens with which to analyze the nexus of wind energy and community economic development. Since the community implications are of central concern, it is crucial to conceptualize community and what it means to engage in community economic development. The term *community* is placed before economic development, serving as a qualifier, reorienting us to the broader spillover effects of economic development beyond the confines of the firm. This review then moves to a critical analysis of top-down versus bottom-up distributed development approaches, with an emphasis on how these systems influence a community's capacity to self-govern. This is important for understanding how we empower communities to recognize economic and environmental justice, crucial in this era of increasing state, market, and environmental disruption. Attention then pivots to perspectives of business ownership models and the connection between ownership and community economic development. The chapter concludes with a discussion of the limited scholarship on cooperatives as community-building institutions.

Community of Place

The community is often conceptualized as identity, solidarity, or place (Flora & Flora, 2008). Some scholars claim community is lost, that it cannot be defined, that it is an amorphous, shifting concept instrumentally useless in an era of globalization (Bauman, 2001). Yet recent literature continues to find analytic efficacy, particularly when embracing a more complex perspective of a community (Flint, Luloff, & Theodori, 2010).

"All of the conceptualizations of the community 'focus on groups of people' exhibiting a 'shared sense of place.' This sense of place forms an ecological perspective, accounting for relationships with the people, cultures, and environments, both natural and built, associated with a particular area" (Flora & Flora, 2008, p. 14). This book adopts the community ecology perspective: "A community may or may not provide the social system through which its members' needs are met. It may or may not provide a sense of identity for its members. What a community does provide is what some sociologists now call locality, a geographically defined place where people interact. The ways that people interact shape the structures and institutions of the locality. Those structures and institutions, in turn, shape the activities of the people who interact" (p. 15).

When a factory closes its doors, the surrounding workforce takes an immediate hit. Businesses that serve the workforce face declining sales and possible closure. When another company seeks to reopen that shuttered factory, the company will probably make requests for subsidies; it will be up to representatives of the workforce, small businesses, and the citizenry to decide if the economic stimulus justifies a diversion of public resources into the bank account of an unknown, unverified community investor.

The community and its space is where an array of collective activity plays out. The structure of the community space and its governance processes are partially determined by broader social-ecological systems. However, the activities of a community and its modes of governance may create healthy friction between these systems. The capabilities of individuals to choose to enter or exit a public debate or action arena and to act collectively when needed are essential prerequisites for democracy (Aligica & Boettke, 2009, pp. 24–25).[1] Communities then provide a critical point of access into society at large and mechanisms of resource governance. Therefore, the aggregated governance and public entrepreneurial behavior of communities and their actors may shape social-ecological systems, just as these systems shape communities.

Tocqueville viewed the community as both a laboratory and school of democracy, where citizens gained the requisite skills to engage in the art and science of association (Ostrom, 1999). Individuals and their communities become more adaptable and robust, exhibiting enhanced tendencies to mobilize ad hoc to address vexing social problems (Poteete et al., 2010) such as climate change (Ostrom, 2014). The inclusion of citizen- and community-level voices also provides for the dissemination of tacit and local knowledge to political leaders and administrators, enhancing the provision of public goods and services. From Tocqueville's vantage point, community-level democracy contributes to the robustness and well-being of state- and national-level democracy (Tocqueville, 2006) by establishing and sustaining self-governing capabilities of the broader polity. From this perspective, the community is at the foundation of the scaffolding of a self-governing democratic order, educating and training citizens to be active in governance while innovating how we do democracy through practice, reflection, and implementation.

In an era of Trumpism, the failures of public administrators, policy makers, and scholars to understand the role of community, civil society, and markets and state—the central concern of Vincent Ostrom's *The Intellectual Crisis in American*

Public Administration (1989)—are more important than ever (Aligica & Boettke, 2009; Ostrom, 1997). The community is not solely the subject of state and market forces but is necessarily a component in an amalgamation of institutions are not subordinate to just economics or coercive authority (Flora & Flora, 2008, p. 14). A community is collective action, people working spontaneously or intentionally on an as-needed basis to address their common concerns (Stedman, Lee, Brasier, Weigle, & Higdon, 2009). The concern for community and individual well-being and its linkage to broader social order are overlooked in our contemporary political discourse on macrostructures such as "the state" and "the market" and by individuals tasked with expert, elite leadership. Community governance is thrown out, impaired by varying degrees of social engineering from on high.

Community Governance

A major economic player comes to town, seeking lavish subsidies in return for job creation and small business opportunities. Community members come together in official (government) and voluntary (nonprofit or oppositional coalitions) settings to weigh the costs and benefits and whether or not to divert significant public resources into the hands of outside private interests. Will their decisions result in an economic boom or a catastrophic bust?

Community governance is "the set of small group social interactions that, with market and state, determine economic outcomes" (Bowles & Gintis, 2002, p. F419). Community governance "is an amalgam of specific practices that make the difference between stagnating and flourishing communities" (Stark, 2007, n.p.). Community governance provides an engagement process for "citizen participation beyond voting" (Gaventa, 2002, p. 29) in which citizens are viewed not merely as electors or consumers but as decision makers and citizen-administrators capable of responding to opportunities and challenges (Flora & Flora, 2008).

Community governance is where culture arises, and the tendencies of community and broader social systems are observable. The community orientations of the actors, their relationships, and their attributes position them to work beyond their narrow interests for the public benefit (Flora & Flora, 2008; Zacharakis & Flora, 2005). Necessarily, this results in vying for influence, struggles for power, and attempts at varying forms of collective action and governance. Vincent Ostrom referred to this process of contestation as "the active and respectful engagement of individuals in proposing and evaluating alternative responses to policy issues, which, since it enables participants to learn more about each other and to develop shared understandings about the problem at hand, is a foundational requirement for effective and sustainable governance" (McGinnis, 2011a, p. 174).

A discussion of community and governance draws us to the issue of social structure. Scholars of social structure are necessarily interested in how the structure of these processes influences an actor's capacity for action (Sharp, 2001, p. 422; Small, 2009). Simply because community governance exists does not mean it is participative, equitable, or just. The focal community governance arena may operate in such a manner that encourages or discourages individual participation (Martinez, 2009, p. 3). Individual agency is neither isolated nor absolute but shaped by individual capability, social interconnectedness, and access to resources (Burt, 1995; Freeman, 2004; Knoke & Yang, 2008), having significant ramifications regarding control and power (Lukes, 2004; Scott, 2000).

Social network research has consistently demonstrated that individuals and community actors with a balance of strong (family and close friends) and weak (work acquaintances, civic associations) social connections have enhanced access to socially mediated resources. Granovetter's (1973) "strength of weak ties" hypothesis posits that an individual's weak ties may provide access to a broader array of resources than an individual's strong ties. Individuals with numerous weak ties tend to be more prosperous, live longer lives, be happier, and be more prone to contribute back to the community (Christakis & Fowler, 2009). It stands to reason, then, that atomization may harm community well-being by eroding weak ties among community members.

Freudenburg's (1986) research of a rapidly changing boomtown assessed how an influx of job seekers moved to the community for new opportunities at the recently built power plant, resulting in divisions within community networks. The division diminished long-standing residents' access to community resources, personal network density (the average proportion of people in a town known by residents), community trust, and capacity for collective action, and individuals erected barriers based on cliques.

Network disruptions may offer new opportunities as well. Burt's (1995) expansion of Granovetter's weak-ties hypothesis builds a theory around how actors gain advantages by connecting weak, delinked, or "brokered" linkages to other networks (structural holes). Knowing this leads to conclusion that, for a robust democratic community governance structure to endure, linkages across a diverse array of networks and individuals are vital for the livelihoods of everyday people.

Perhaps disruptions can be better controlled through robust community governance. Forms of community governance create predictable patterns and social order, pathways by which individuals may voice common concerns and address common concerns collectively. A community's governance structure can catalyze adaptability via interactive processes connecting diverse actors with multiple interests. Inclusive communities could better welcome in-migrants and sustain preexisting community dwellers, cultivating reciprocal relationships and connectivity across networks.

The challenge for researchers and community economic developers is to embrace the uncertainty that there are no one-size-fits-all models for governing a community. The empirical work demonstrates that, for conditions relevant for robust community well-being to arise, governance structures must allow for inclusive, redundant, diverse social institutions so that optimal institutional types may develop and adapt to unique, local socioecological features (Poteete et al., 2010). Elinor Ostrom, in her seminal *Governing the Commons* (1990), finds that communities bringing diverse stakeholder groups together to govern and comanage natural resources are often more resilient to system shocks. Centrally managed municipal water systems, for example, taking dictates from a state-level authority, may be disconnected from the needs of local resource users. This disconnection could result in misuse and degradation of the system. However, a more collaborative approach that engages consumers, industrial users, and scientists in shared decision making may provide more socially equitable, environmentally sound outcomes informed by community stakeholders.

Communities with vibrant, active governance serve as incubators of democracy by giving actors the ability to observe how governance works, participate in the process, and work with others who also desire to be heard and pull the decision-making levers. Participative communities are more adaptable to change, and system shocks and actors are more likely equipped with the appreciable capacity to steer change in a positive direction, enhancing agency and local control (Emery & Flora, 2006; Shragge & Toye, 2006). Therefore, social structure matters, but so too does the manner in which collective action occurs (process) within that structure and on whom those actions bestow benefits.

While the Tocquevillian perspective positions community and community-level governance as a requisite for democracy, one should not fall into the trap that devolution of public goods and services to the community level is necessarily ideal or optimal. Community control as the panacea is a misunderstanding of dominant paradigms of community economic development (self-governance and self-help). "By failing to be vigilant, analytic and focused on our principles and purpose, we leave ourselves open to manipulation and dilution" (Ledwith & Campling, 2005, p. 4). A perverse romanticization of the term *community* has been used by state actors to shirk their responsibilities by playing on false narratives manipulating the place of community in providing for the general welfare. There are those public goods, such as health insurance, that are economically optimized at scale and best left to well-resourced, expert leadership. The state can indeed be a partner for enhancing community well-being.

The ongoing "small state" debate posits communities as both society's problem and solution (Ledwith & Campling, 2005, p. 26). An over-romanticization of community self-help capabilities combined with the deprioritization of state provision of public goods and services has devolved public goods and services provisioning from the state to the community. "We witness the full might of the state in Co-opting civil society as an alternative provider of state services, eroding rights and magnifying" community and individual "responsibilities still further" (p. 28). We then create a situation in which we task resource-constrained communities with enormous duties beyond their capabilities, previously performed in an aggregated manner by well-resourced state actors. The inability to critically analyze and collaborate to address social problems allows powerful forces to continue unabated, exacerbating social fissures. "Large inequalit[y] . . . weakens community life, reduces trust and increases violence" (Wilkinson & Pickett, 2009, p. 45). The results have been disastrous.

We can observe these processes at play in the Reagan-era drug war which disproportionately incarcerated black men despite similar rates of criminal behavior by whites (Ledwith & Campling, 2005). As national-level safeguards receded, "states' rights" policies provided a space to codify local and regional racial grievances into law; black voices were ostracized in favor of a white cultural perspective that inaccurately perceived black bodies as the primary source of crime waves and general criminal behavior. When left unchecked, this extralocal process creates a self-regenerating process of community atomization, decay, and desperation. There remains a place for economies of scale, state-level decision making, and expert administration in service provision that, if rejected, could result in communities that are unduly handicapped or allow for undemocratic elements to arise. The next section examines how to direct communities toward the fulfillment of an empowered self-governing polity.

Community Economic Development

I recall as a child the fervor in Mattoon, Illinois, welcoming a new bagel factory to town. The city's success in lassoing a major employer was our success! The city of Mattoon started Bagelfest to celebrate the hundreds of newly created factory jobs, a celebration of the bagel factory that to this day draws in tens of thousands for free bagels and up-and-coming country music stars. To this day, Mattoon still celebrates the bagel, even though the factory employs a fraction of the original number of workers. Should Mattoon be revering the bagel? In the end, what did the bagel do for Mattoon and those dependent upon the jobs provided?

One should not conflate the mere act of participating in a community and its governance with the practice of community economic development. With an uncritical lens community economic development lapses into one-off token projects that do more for feelings of edification and little for broader economic justice (Ledwith & Campling, 2005, p. 23). Under such conditions, we see processes such as structural racism arise as one group (whites) secures a dominant governance position above another (nonwhites). Radical concepts such as participation are appropriated, used to devolve public goods and services provision to specific communities and individuals. Within the context of community development, these interests may be perfectly qualified, usurped, or worse, corrupted out of self-seeking interests. The "social, political, and economic macro-structure cannot be side-stepped" (Ledwith & Campling, 2005, p. 29), but it can be accessed and reshaped.

Community economic development entails more than strengthening the existing community structure. A critical community development orients the developer toward a process for social change through the social structure, as opposed to atomized projects. Community development is itself the practice of identifying and attempting to change the actual social structure underlying the community itself (Wilkinson, 1991).

Participating in the local Fourth of July parade may maintain existing social order, but it changes little. While participating in a mainstay community event may not change the social structure, taking on the community's most prominent landlord to provide affordable rents and better tenant services might. A landlord clique could have control over significant community assets, but by providing underrepresented voices a seat at the policy-making table, we may alter the community's social structure. These voices may then harness a seat of power, providing concrete material and social gains.

The International Association for Community Development (2012, n.p.) defines *community development* as "a practice-based profession and an academic discipline that promotes participative democracy, sustainable development, rights, economic opportunity, equality and social justice, through the organisation, education, and empowerment of people within their communities, whether these be of locality, identity or interest, in urban and rural settings."

Community economic development typically falls under three approaches: (a) technical assistance, which provides top-down, expert-driven services; (b) conflict-oriented (Alinsky, 1971), which emphasizes a power redistribution approach by way of professional organizers working with community members to identify a target, shaming the target's practices and threatening the target's franchise to get concessions and access to power; and (c) self-help, which accentuates self-sustainability through the identification of community assets that can be accessed, harnessed, and deployed for the collective benefit (Flora & Flora, 2008, p. 44).

Community economic development requires a critical understanding of who wields control and influence—power—over vital resources. A "radical practice" has tendencies of all three approaches but is broader in scope. A radical practice has at its base a transformative agenda. Part of the agenda is to bring about social change grounded in a fair, just, and sustainable world, locating "the roots of inequality in the structure and processes of society, not in personal or community pathology" (Ledwith & Campling, 2005, p. 14). Community development is rooted in the lives of everyday people, based on a foundational

> process of empowerment and participation. Empowerment involves a form of critical education that encourages people to question their reality. This empowerment orientation is a basis of collective action built on principles of participatory democracy. In the process of action and reflection, community development grows through a diversity of local projects that address issues faced by people in a community. Through campaigns, networks, alliances, and movement for change, this action develops a local/global reach that aims to transform the structures of oppression that diminish local lives. (pp. 2–3)

The inclusion of empowerment and participation of the marginalized is central to community economic development. "Empowerment means changing the relationship between the rich and the poor, not the false option of 'breaking it.' . . . Empowerment cannot be depoliticized, . . . involving a process of critical consciousness"— "a stage of consciousness" linking critical thought and action "needed . . . to collectively act relative to the wider context of power from local to global"—"as a route to autonomous action." This route is "not an alternative solution to the redistribution of unequally divided resources" (Ledwith & Campling, 2005, p. 29). "People are encouraged to ask thought-provoking questions and 'to question answers rather than merely to answer question." The process of problematizing "exposes structures of power and the way these impact on personal lives," challenging "pathologizing theories that lay blame and responsibility at the feet of victims of injustice" (p. 10). In doing so, the marginalized are better equipped to address structural inequalities, to move social systems in a more equitable direction by becoming craftspersons or designers of their realities as opposed to passive subjects. The citizen could be exposed to and serve in multiple roles, becoming a developer, organizer, planner, analyst, academician, and so forth.

Community economic development is then more explicit in its ends than community governance. Community governance serves "as both a laboratory and school of democracy, where citizens gained the requisite skills to engage in the art and science of association," community governance is frequently delinked from an understanding of social order and pathways of oppression (Ledwith & Campling, 2005, p. 10). Community economic development illuminates the individual to see power and structure, to alter their reality collectively and in solidarity with others (p. 10). The practice of community economic development more explicitly "calls for a unity of theory and practice (praxis). Informed by antidiscriminatory analyses, and in a symbiotic process of action and reflection, critical analysis deepens with practical experience. In this way, action generates theory, and action, in turn, becomes more critical through analysis. Inspired by a vision of a more just, equitable and sustainable work, this aspiration is not only a possibility, but a necessity" (p. 3).

We move away from the deification of business leaders and long-worn political dynasties. Community economic development praxis "locates the silenced stories of those who are marginalized and excluded" (Ledwith & Campling, 2005, p. 9). "It builds on a grassroots community activism, developing projects that are based on a sustainable living," accessible "local economics and human values, but reaches out in alliance to change the root causes that give rise to structural discrimination" (p. 3). In this manner, community economic development "occupies a contested space between top-down and bottom-up" (p. 3), adding a level of intentionality to community governance in the form of justice and orientation toward the greater social order beyond the community.

The orientation of a lead developer or organizer is essential. The critical community economic developer is not just a facilitator but also a teacher and, importantly, a fellow student who also engages in praxis. Not only does the developer have much to teach, but the students also become teachers by assisting the boundedly rational teacher-developer see the unseen and otherwise obscured (Ledwith & Campling, 2005, p. 2).[2] The community economic developer seeks to understand social order and to change it in solidarity with those most impacted. Ledwith and Campling (2005, p. 11) suggest that a community developer continually reassess four overarching questions in their praxis to maintain focus on a given community development endeavor:

- How does an antidiscriminatory analysis influence your practice?
- In what ways do national and global changes impact the diverse lives of local people?
- What evidence is there that your practice contributes to change for social justice and environmental justice?
- How can you extend collective action beyond the boundaries of the community to build national, international, and global alliances?

A community nests within the broader social order. Take renewable energy policy, which devolves—or delegates—policy implementation to a private entity or firm. The firm is itself a creature of the macrosocial order, a legal "corporate actor" engaged in policy implementation constructed in "higher-level" action situations (McGinnis, 2011b, p. 9) outside of the immediate reach of the community. This is true for any corporate legal entity, such as nonprofits, limited liability corporations, and cooperative enterprises. The implications are significant for community members seeking to engage with such entities. Not only do engagement mechanisms exist locally, but they exist at other levels of government and within the economy.

Awareness of institutional diversity within a social order is critically important. Institutional diversity is needed to provide the requisite variety of solutions that properly align or fit a given social problem and the collective responses needed to fit the various time-and-space contingencies found in different settings, "to make effective use of local knowledge" (p. 5). A community comprises mostly investor-owned firms might be presumed to take on many of the characteristics and traits of those firms due to the weight of their influence. From a rules perspective, the collective interests of such firms may result in collusion on public policy prioritizing their interests—shielded behind the cloak of legalese—at the expense of other forms of enterprise. A community comprising one specific firm type (e.g., IOUs) may face entrepreneurial barriers to forming diverse types of firms (e.g., cooperatives). The lack of an observable alternative to the status quo may degrade the citizen imagination.

Attention now turns toward the matter of institutional design and intentionality.

The Role of Intentionality

Monocentric systems limit voice, choice, alternative institutional arrangements, community robustness, and incubators of democracy. However, even an oppressive social order is never wholly monocentric and often contains numerous pathways available for those individuals seeking to engage in community economic development.

This diffuse system of alternatives that may compete, cooperate, or coexist within a social system is referred to as *polycentricity* (Aligica & Boettke, 2009, p. 3), "a pattern of organization where many independent elements are capable of mutual adjustment for ordering their relationships with one another within a general system of rules" (McGinnis, 2011a, p. 175). Polycentricity involves multiple authorities with varying degrees of responsibility and control over any number of adjacent or overlapping jurisdictions. Recognizing and embracing the complexity inherent in all social order allows the institutional analyst, community economic developer, or planner to observe and strategize around the rich tapestry of institutional interactions, the array of locales able to undertake crucial decisions and entrepreneurial endeavors. "No single center of authority is responsible for coordinating all relationships in such a 'public economy.' Market-like mechanisms can develop competitive pressures that tend to generate higher efficiency than can be gained by enterprises organized as exclusive monopolies and managed by elaborate hierarchies of officials" (p. 7). In other words, when we remove unnecessary barriers to opportune structures and when certain conditions exist, people are able to act collectively to find common solutions to social dilemmas that are outside of just "the state" and "the market."

Polycentricity and pluralism are not sufficient for a radical community economic development. A pluralistic social order is one that provides options, and as with all social orders, it may or may not be purposive (exhibiting a *telos*, or purpose or end goal, as constructed by given actors and institutions). The design and intent of a polycentric social order—and, by extension, a radical community economic development—must not be ignored if we desire a more just social order. "The key distinguishing feature of a radical, as opposed to pluralist practice, is the political consciousness that united people in collective action beyond the boundaries of neighbourhood to engage in wider structural change" (McGinnis, 2011a, p. 197). Polycentricity is not the end; it is instead a desirable requisite for robust collective action. Just because a polycentric system lacks a central authority does not mean it is incapable of overcoming these obstacles and ordering itself (Poteete et al., 2010, pp. 39–41).

Take credit unions, for example. The year 2008 saw the inevitable result of a market economy reliant upon for-profit banks. These banks wielded inordinate political power, leveraged to enable predatory behavior in the mortgage markets. As the mortgage markets collapsed, so to did the banks, leaving the economy in shambles. Except the American credit union sector remained robust, with a failure rate much lower than the for-profit banks. Where the for-profit banks failed, credit unions continued to be liquid, providing consumer mortgage products, whereas the banks required an enormous public subsidy to return to basic functionality.

A system can be structured in such a manner as to prevent individual and collective action leading to atomization or centralized power. Credit unions were not durable because they were sovereign and isolated. They were embedded in the banking industry. Their conservative, risk-adverse nature strengthened the credit unions, and their federated structure further shielded them from the excesses of the for-profit banks.

Institutions provide an access point into the broader community and broader society. "Collective action for change has to follow through from local to structural levels in order to make a sustainable difference" (Ledwith & Campling, 2005, p. 14). Leaving development to spontaneity, the "invisible hand," means we allow the most advantaged among us to take the helm. So, while policy may claim adherence to spontaneity and free market competition, the observed reality is otherwise.

If we are to believe Klein's (2007) disaster capitalism hypothesis—that private capital exploits natural disasters and civic crises to the detriment of the public good—then it becomes clear that many of our current economic regimes are not purely the result of spontaneous free markets. Profit-oriented interests shape the political discourse through the media and policy. The alignment of profit interests with public policy sets forth an incentive system that handsomely rewards those with the capability to write the law at the expense of the many. Sure, polycentrism and institutional diversity may exist, but the system may progress toward centralized benefit.

If we structure the broader social order on a principle that profit-seeking entrepreneurship is the most virtuous, then it follows that institutions aligned with these interests will rise to the top and may choke out alternatives. It is indeed true that a significant number of profit-seeking interests contribute to charitable giving and the public good, but a social welfare system based on privileged charitable giving is limited. Great, the TOMS brand, with every purchase made, provides a kid in need with a pair of shoes, but the kid remains in need, albeit with a new pair of shoes. These well-meaning attempts at easing the excesses of the capitalist enterprise are frequently "ameliorative, making life just a little bit better around the edges, but not stemming the flow of discriminatory experiences that create some lives as more privileged than others" (Ledwith & Campling, 2005, p. 14).

Absent certain pretexts, functions like basic interaction, information sharing, trust, contractual obligations, and sustained collective action are unlikely to endure (Poteete et al., 2010, pp. 100–101). We distort the features of an institution when their ideal-type design is interpreted through the lens of public policy and operationalized in a given manner by the institution's stakeholders. The laws and statutes overseeing community-based organizations in many countries may serve an insufficient guidepost for their stewardship. Indeed, under U.S. state statutes and tax law, many community-based organizations are relegated and structured in such a manner to incentivize behavior similar to that of the investor-owned firm (a homogenizing effect). We should not be surprised when we see cooperative institutions behaving more like profit-oriented firms than an organization with a deep commitment to its members' and community's well-being. This is why it is critically important for individuals to be aware of the legally constructed social order, the ideal type of a given institutional model, and the governance processes congruent to the institution's desired outcomes (and, ideally, the legal, social order should recognize the variations in institutional form). Barring such awareness, institutional distinctiveness is shrouded, and systemic innovation and adaptation become more difficult, more prone to the pitfalls of monocropping.

The lesson is that community economic development is not empowerment occurring at the fringes, but one requiring a hands-on, purposive orientation, aimed at altering social system structure from below. "Empowerment is a transformative concept but without a critical analysis it is all too often applied naively to confidence and self-esteem at a personal level, within a paradigm of social pathology, a purpose that is usually associated with personal responsibility for lifting oneself out of poverty, overlooking structural analyses of inequality" (Ledwith & Campling, 2005, p. 13).

Open systems of governance allow for innovations that could enhance fundamental efficiencies or promote a radical community economic development. Within those systems, alternative institutional arrangements for collective action (e.g., social-purpose businesses, nonprofits) are more apt to thrive within a foundation of polycentricity (Aligica & Tarko, 2012), where openness provides for greater experimentation and entrepreneurship.

Central to this study is the idea that institutional form and ownership matter. Within that, a specific type of institution, the cooperative, is itself an overlooked implementation strategy for community economic development (Zeuli & Radel, 2005). It is necessary to discuss how differences in institutional form, defined by the mode of ownership, interact with community economic development.

Institutional Ownership, Stakeholder Control, and Community Economic Development

What difference does the institutional diversity and composition of a community make? Picture two midsize communities of 100,000 residents. The only thing that you know about the community-level economies is that one community is composed solely of investor-owned firms, whereas the other is a mix of nonprofits, cooperatives, and investor-owned firms. If I were to ask you which community composition has the most significant positive social impact, what would you say?

I speculate that, regardless of whether you, the reader, lean right or left, Democrat or Republican, you will have an opinion as to which community promotes a more prosperous society. What this indicates is that institutional design and composition do indeed matter, with ramifications for community and social well-being.

Institutions are themselves artifacts or constructs shaped by the actions of individuals and society. These institutions reinforce behaviors and mores "more essential to the continued development of democracy than the details of its legal system or its physical location" (McGinnis, 2011a, p. 196). The features of an institution determining their structure, whom they serve, and their obligations and restrictions, are informed by the *institutional logics*, defined as the socially constructed, historical pattern of material practices, assumptions, values, beliefs, and rules by which individuals produce and reproduce their material subsistence, organize time and space and provide meaning to their social reality. . . . Institutional logics are both material and symbolic—they provide the formal and informal rules of action, interaction, and interpretation that guide and constrain decision-makers in accomplishing the organization's tasks and in obtaining social status, credits, penalties, and rewards in the process. . . . These rules constitute a set of assumptions and values, usually implicit, about how to interpret organizational reality, what constitutes appropriate behavior, and how to succeed. (Thornton & Ocasio, 1999, p. 804)

Institutions exist for a purpose or end goal (telos), in service to specific stakeholders. Ownership specifies which stakeholder class the institution is beholden to, informing how institutions function, and depending on the ownership class, this may enhance or impinge community economic development outcomes and policy objectives. General Electric exists to provide a return on shareholder investment; community economic development is an unintended spillover. The consumer ownership of an electric utility monopoly orients management toward meeting consumer needs. Replacing the profit motive (no shareholder exists demanding a dividend) with a member needs orientation then reduces the cost of delivering goods and services while enhancing consumer-oriented governance, minimizing the need for external monitors or government regulators to keep institutional excesses and managerial self-dealing at bay (Hansmann, 2000, p. 6).

We turn now to the theory of ownership to speculate on how various types of institutions may interact with local communities and the broader social order, going beyond just the Bloomington School perspective, pulling from the broader field of New Institutionalism. While an array of institutions exists, this comparative analysis focuses only on two: cooperatives and investor-owned firms. Unlike the sole proprietorship or the nonprofit, investor-owned firms and cooperatives are collectively owned, but with different stakeholders as owners (monied interests vs. people interests) and decidedly different perspectives on their social responsibility. The diversity and heterogeneity of owner preferences must be accounted for through collective choice mechanisms, which themselves provide valuable insights for community governance and our broader democratic social order. "By studying the structure of ownership in private firms we gain a strikingly strong perspective on the relative virtue of politics and markets in governing social activity—a question that has been at the center of Western political and economic debate over much of the past two centuries" (p. 4).

Theory of Ownership

Ownership is the formal control of a firm, with the owner being the rightful claimant to the firm's residuals (Hansmann, 2000). Bundling control rights with residuals incentivize the ownership class of stakeholders to participate in governance to maximize returns. While owners of collectively held institutions like investor-owned organizations formally control the firm, they do not adequately control it (outside of uncommon circumstance). The delegation of control to management is a function of what is called the separation of ownership and control. The granting of substantial autonomy to management presumes that the complexity exhibited in the scale and scope of day-to-day activities of any given firm — as well as the dispersion and heterogeneity of owner interests — is best left to expert managers to be harnessed for the benefit of the ownership.

Laws, ownership, and norms all converge to dictate the governance of a firm. Becht, Bolton, and Röell (2003, p. i) define corporate governance as being "concerned with the resolution of collective action problems among dispersed investors" or specified stakeholder class "and reconciliation of conflicts of interest between various corporate claimholders." This is then translated into "the rules and practices that govern the relationship between the managers and shareholders of corporations, as well as stakeholders like employees and creditors" (Jesover & Kirkpatrick, 2005, p. 2). Governance further specifies the processes and procedures for making decisions not defined in law or statute.

A major corporation such as Walmart is owned by thousands of shareholders with a diverse array of personal preferences, running the gamut from self-seeking to altruistic. The politics of the firm "have a critical bearing on the patterns of ownership we observe and the ways in which firms are structured internally" (Hansmann, 2000, p. 2). We see this in governance. The delegation of control of the collective to board and management simplifies the complexity of owner preferences in collectively owned firms. While many forms of governance exist (Sherwood & Taylor, 2014), principal-agent dominates. The principal-agent model persists due in part to the widely held perception that the delegation of control by the collective (the principal) to the board and management (the agent) accounts for many costs best accounted for by expert administrators (Ghoshal, 2005).

The Intersection of Law and Institutional Behavior

Both the investor-owned firm and the cooperative are designed to operate on behalf of their ownership stakeholder class. However, cooperatives diverge substantially in their governance, which then informs their performance, as well as their impact on community economic development.

The cooperative is predicated on the ICA cooperative principles and values (CPs; see Table I.1). Cooperatives are owned collectively by the community of member-owners, the users of the firm. Nonusers may own a locally owned firm (e.g., a landlord). In a co-op, the weight of ownership is based on not the share of investment but individual personhood: cooperatives provide each member one vote, no more and no less. Ownership also confers governance rights, allowing the community of member-owners to shape the behavior of the firm. Additionally, cooperative member-owners are guaranteed by law to participate in the economic features of the co-op (Cook, 1994), meaning that, if a cooperative generates margins (profits), member-owners are entitled to receive a dividend in proportion to their patronage. Taken together, these embedded features of the cooperative, reinforced by legal statute, engender feelings of ownership, trust, reciprocity, and solidarity, features not legally required of investor-owned or locally owned enterprise (though they may do so voluntarily) (Cook, 1994).

Under the patchwork of inconsistent, underdeveloped cooperative statutes and regulations, cooperatives are disadvantaged relative to investor-owned firms (Hansmann, 2000). According to Cook (1994, p. 45), U.S. law codifies three of the seven CPs (democratic control, service at cost, and limited return on equity). The U.S. approach to co-op law is uncommon among Western democracies: most European countries have codified all seven CPs (Cracogna, Fici, & Henrÿ, 2013). In the United States, the four remaining CPs are left to the discretion of the cooperative's agents to implement. The legal guidance emphasizes economic well-being with no substantive directive toward the social benefit features of educating the membership, coordinating with other cooperatives, or exhibiting concern for community. This severely impairs the credibility of the cooperative in committing to the development of solidarity (bonding social capital) among its membership, community, and other cooperatives within broader society. Functionally, U.S. law cuts against the ideal type model of the cooperative, orienting cooperatives toward economic or financialized outcomes. Cooperatives cannot be presumed to be community-building institutions without an explicit statement of commitment by each cooperative to the CPs or the force of law. The National Rural Electric Cooperative Association (NRECA) and its entire electric cooperative membership make such a commitment.

The law's relative silence on governance has the effect of handicapping the cooperative's efficiencies. Democratic control tends to be interpreted by board and management as a regularly scheduled vote for members of the board, adhering to the dominant business perspective of separating ownership from control. The restriction of patron participation in essential governance and management decision making undermines theoretic efficiencies of the cooperative. Information asymmetries persist by limiting the flow of member-owner preferences to the agents, further constraining the generation of social capital (Hansmann, 2000, p. 5). According to Taylor (2015, p. 149), effective governance is diminished

> to reductive input-output functionalism, with organizational control mechanisms placed predominantly in the board-management relationship. This is opposed to the engagement of a broader stakeholder base that contributes by providing value in the form of critical information and policy solutions . . . having the impact of putting a stop to any manner of radical social reconfiguration by blocking pathways for broad-based public participation. . . . This is highly problematic if we are to believe that the third sector is to play a significant role in broader policy deliberations within a pluralistic democracy.

While all cooperatives must provide each member-owner the right to vote on crucial issues, the bylaws of the cooperative may reduce the array of issues decided by the member-ownership. Even then, voting as the sole instrument of expressing collective voice may unintentionally alienate marginalized groups within the collective ownership, limiting the mechanisms for influence by diluting the voice of the outlying member-owners.

Despite the legal and social norms privileging principal agent governance, many cooperatives—and investor-owned firms—practice modes of governance that to varying degrees engage the ownership and other patron classes such as consumers, workers, and producers (Laloux, 2015). There is emergent literature on organizations successfully using an array of governance models, such as distributed leadership (Spillane, 2012) and multistakeholder governance (Bäckstrand, 2006). According to Hansmann (2000), if legal frameworks were to better recognize the unconventional modes of governance applicable to alternative institutions like cooperatives, these institutions could gain significant efficiencies above and beyond their investor-owned counterparts. Identifying where these alternative modes of governance are being practiced, how they came to be justified, and their relation to organizational performance remains an important topic for further study.

The member-ownership design of cooperatives, as well as government securities regulations, places cooperatives at a capital disadvantage relative to investor-owned firms. Investor-owned firms may seek unlimited dollars through stock offerings by any investor. Cooperatives are confined to investment by their patron member-owners, though this barrier may be partially overcome by debt servicing (owners of a firm are not necessarily the providers of capital). That said, the practice of seeking member-provided capital in the form of member loans is another useful tool. Member provided capital is useful for (a) engaging member-owner engagement, (b) enhancing community wealth and social impact by distributing the loan interest back to the community as opposed to a bank or external investor, and (c) developing endogenized sources of capital for future community-building projects.

Cooperatives should exhibit additional efficiencies through the internalization of their social costs. The participation by the cooperative's member-ownership diminishes these information asymmetries through the revelation of patron preferences to the board and management. This augmented information flow mitigates the public-service paradox, providing previously obscured knowledge to management. The active participation of the general member-ownership further reduces monitoring costs.

These efficiencies tied to preferences allow the cooperative firm greater agility and responsiveness to changing member needs and market circumstances. The investor-owned firm, however, must bear these costs because they do not pull upon their ownership base in the same manner. Patrons of an investor-owned firm have little incentive to provide in-depth information that could enhance firm performance.

Regulatory policy further advantages the investor-owned firm. Governmental consumer protection agencies provide an advantage to many investor-owned firms by controlling for their excesses. Monopoly electric utility providers fall under regulatory compacts and are granted monopoly rights under tight regulation that curbs exploitive rent-seeking behavior by management. Investor-owned firms are in a sense "saved from lesser angels," providing them with the enhanced competitiveness that may "ultimately displace cooperative . . . firms" (Hansmann, 2000, p. 5). One such mechanism is regulatory price setting. Many public utility commissions allow electric utilities 9–13 percent margins, an extraordinarily high rate of return on a commodity product. Self-regulating cooperatives, however, frequently operate on a 1–4 percent margin. While there are substantive impediments to cooperative entrepreneurship, cooperatives continue to persist throughout many industrial sectors.

What, then, is the case for the utilization of cooperatives as part of a broader community economic development strategy?

Institutionalizing Community Economic Development: The Cooperative Business Model

The investor-owned firm may have a great deal of impact on community economic development. The firm may provide jobs and deep investment. There is nothing to keep the investors of the firm from going beyond these ameliorative benefits toward the engagement of a critical community economic development initiative. Indeed, industrialist Edward Filene used his considerable wealth to help fund the growth of the credit union movement in the early 1900s; Filene's legacy encompasses over six thousand credit unions in the United States alone. Simply put, investors could dictate that their firms invest in their host communities, and many do.

However, the benefits of investor-ownership are designed with the imperative of capital (he who has the most money has the most voice—the primacy of capital). The investor-owned firm is wrapped in the institutional logic of markets (Thornton, 2002), whereas the institutional logics of member service or need are the foundation of the cooperative. In the cooperative, the benefits are meant to be conferred to people (one member, one vote—the primacy of the individual). The tangible benefits of economic opportunities stemming from cooperatives—job creation and community cohesion—are apparent (Brennan, 2009; Ki-moon, 2009). But what precisely are the features of a cooperative that gives it a community orientation, uniquely differentiating it from the traditional corporation, and how, then, might cooperatives be used for community economic development?

A cooperative is an inherently political form of institutionalized collective action (Mooney, 2004), a common-property regime that according to the ICA's website is "an autonomous association of persons united voluntarily to meet their common economic, social and cultural needs and aspirations, through a jointly owned and democratically controlled enterprise" (International Co-operative Alliance, 2015, n.p.).[3] The cooperative model of business, like the corporation, is utilized in virtually every industrial sector but operationally functions much differently (Brennan, 2009; Restakis, 2010).

The divergent institutional logics result in many variances, which any policy maker, developer, analyst, or planner should be cognizant of when aligning policy to desired outcomes. Table 1.1 outlines some of the key differential institutional logics between the cooperative and the investor-owned corporation. Take, for instance, the application of the investor-owned or co-op-owned firm in a given policy matter. Whereas the corporation ties control to an investor's share of total investment (one share, one vote), voting rights in a cooperative are equitable (one member, one vote). Equitable voting rights provide for a more representative governance structure. Whereas the largest shareholders typically compose the board of a corporation, the ranks of a cooperative's members elect and serve on the co-op's board. Whereas corporations produce profits (and pay dividends on stocks), cooperatives produce patronage refunds or surplus wealth distributed back to the membership based on the level of patronage, creating an environment of shared prosperity and enhanced individual opportunity.[5] Whereas corporations seek value appropriation and profits (Flora & Flora, 2008, p. 15), not value creation (Santos, 2012), cooperatives generate value in the form of collective wealth stewarded by the member-ownership with guidance from the CPs, which actively acknowledge "the link between cooperatives and the institutionalization of community development"

Table 1.1. Types of corporate and cooperative institutional logics

Characteristic	Investor-owned corporation	Member-owned cooperative
Economic system	Market capitalism and competition	Market share and distribution
Organizational identity	Products or service for profit	Products or service at cost
Legitimacy	Market position of firm	Organizational reputation (member-owners and community)
Authority structure	CEO, shareholder ownership	CEO, member-ownership, community
Mission	Build competitive position	Increase organizational robustness
	Increase profits and cash flow	Meet member service needs
Focus of attention	Resource competition	Member service needs
Strategy	Acquisition growth	Embed with member culture
	Build market channels	Account for market failure
Logic of investment	Capital committed to market return	Capital committed to member well-being
Residuals	Dividends: residuals based on share of total ownership	Patronage: residuals distributed to member-owners based on share of business
Governance	Primacy of capital: weighted by proportion of voting shares	Primacy of the individual: one member, one vote democratic basis; weight is not adjusted for any reason

Source: Adapted from Thornton, Ocasio, and Lounsbury, 2013; (inspired by a similar chart listed in Thornton, 2002, p. 85).

(Zeuli & Radel, 2005, pp. 45–46). Whereas corporations are associated with capital flight, cooperatives are known to root their wealth by investing back in their host communities, closer to the member-owners (p. 50). Whereas corporations build selected capacity among a marginal spectrum of institutional stakeholders, cooperatives generally build community capacity (human, social, political, and financial capital), serving as incubators of democracy and public entrepreneurship, fostering local leadership and educational opportunities for virtually all stakeholders of the cooperative (members, managers, employees, and the broader community) (p. 48). Whereas the investor-owned firm seeks market and political control through competition and consolidation, cooperatives federate, maintaining localized autonomy while providing access to local and broader social order, providing market and political power to what may otherwise be disconnected populations.

Taken together, the features of a cooperative provide a venue for assembly, constructing new purposes for individuals to convene. Such spaces foster social capital (Putnam, 2000) through civic interaction (Small, 2009), creating new resources and capacities for otherwise disconnected, segmented populations, facilitating the building of norms, trust, and relationships (Putnam, 2000; Small, 2009; Tolbert, Irwin, Lyson, & Nucci, 2002). The cooperative business model then serves a pedagogical and cultural reproductive function by instilling self-governing, democratic values in its membership and partner organizations through practice, operations, and training (Emery & Flora, 2006).

The underlying CPs lay the foundation for organizations with a strong social tilt to use market-like features (i.e., revenue streams from services rendered). The social tilt and market features sustain prolonged struggle against the fluctuations in resources that other organizations (e.g., nonprofits) face most acutely from state and market forces (Brennan, 2009, p. 2; Mooney, 2004).

The on-paper orientation of the cooperative business model parallels Elinor Ostrom's design principles (ODPs; see Table 1.2) for enduring sustained collective action through robust institutional arrangements (Poteete et al., 2010, pp. 100–101). In this manner, cooperatives should mitigate disempowerment, alienation, and dependency-building mechanisms of monocentric systems and enhance the potential for the development of polycentric self-sustaining, self-governing institutions with a community economic development bent.

One must temper the radical expectations of cooperatives. As discussed, numerous legal, normative, and cultural constraints exist, but impediments exist in other arenas. The academic literature narrowly defines and treats cooperatives as a "business first and foremost," run predominantly for the direct benefit of the membership (Zeuli & Radel, 2005, p. 48). Therein lie the tension and the opportunity: if the collective membership is not aware of and does not demand a radical community economic development orientation, then the cooperative will not necessarily tilt in such a direction. Even if the member-owners are aware, they may not be so inclined. From this perspective, a cooperative has much more in common with a community governance institution than with a definitive form of radical community economic development. Many cooperatives, consumer owned specifically, have open membership, meaning that anyone in the community may participate. In this manner, the more cooperatives we have in a community and the broader social order, the more we afford opportunities to the broader polity for active participation in the political and economic system at the local, state, national, and global levels. In this

Table 1.2. Ostrom design principles

Principle	Description
ODP 1A: User boundaries	Boundaries between legitimate users and nonusers must be clearly defined.
ODP 1B: Resource boundaries	Clear boundaries are present that define a resource system and separate it from the larger biophysical environment.
ODP 2A: Congruence with local conditions	Appropriation and provision rules are congruent with local social and environmental conditions.
ODP 2B: Appropriation and provision	The benefits obtained by users from a common-pool resource, as determined by appropriation rules, are proportional to the amount of inputs required in the form of labor, material, or money, as determined by provision rules.
ODP 3: Collective-choice arrangements	Most individuals affected by the operational rules can participate in modifying the operational rules.
ODP 4A: Monitoring users	Monitors who are accountable to the users monitor the appropriation and provision levels of the users.
ODP 4B: Monitoring the resource	Monitors who are accountable to the users monitor the condition of the resource.
ODP 5: Graduated sanctions	Appropriators who violate operational rules are likely to be assessed graduated sanctions (depending on the seriousness and the context of the offense) by other appropriators, by officials accountable to the appropriators, or by both.
ODP 6: Conflict-resolution mechanisms	Appropriators and their officials have rapid access to low-cost local arenas to resolve conflicts among appropriators or between appropriators and officials.
ODP 7: Minimal recognition of rights to organize	The rights of appropriators to devise their own institutions are not challenged by external governmental authorities.
ODP 8: Nested enterprises	Appropriation, provision, monitoring, enforcement, conflict resolution, and governance activities are organized in multiple layers of nested enterprises.

sense, cooperatives are latent community economic development capacity, waiting to be harnessed. It is up to the member-ownership to make it so.

No doubt some cooperative models that have arisen in recent times present challenges for community economic development purposes (e.g., the outdoor gear retail cooperative REI, whose national scale makes localized member passage into governance unduly complicated), complicating place-based development by sepa-rating member-owners over a considerable distance. The explicitly stated purpose of the cooperative and the actors operating it coalesces to determine the member and community orientation of the cooperative (bottom-up, top-down, peripheral, or radical; Zeuli & Radel, 2005, p. 48), meaning the community aspects of cooper-atives are certainly not predetermined. Consider that some communities without the necessary capacity to act collectively and sustain a cooperative nonetheless start a new one (p. 51). So, even though U.S. cooperatives predominantly adhere to the CPs, they typically specialize in the provision of a single service, unlike their social cooperative counterparts in Great Britain, Sweden, Canada, and Italy (Fazzi, 2011; Zeuli & Radel, 2005). This specialization potentially limits their sociocommunity reach and community staying power. It may be that the pervasiveness of the corpo-rate system has instilled a dominant logic in cooperative organizational governance and operations, weakening the influence of the CPs. In this manner, cooperatives, like community governance, cannot be presumed to deliver on a radical commu-nity economic development. It is up to the board, management, and members to make it so. However, advocates must also be aware of these facts, lest corporate logics undermine the initial purpose of their co-op's entrepreneurship.

Advancing Understanding of the Development Impacts of Institutional Ownership Models

How can an understanding of the ownership and institutional logics provide in-sight into how firms interact with their communities? There is a presumption that cooperatives are local businesses, often situated within spatial proximity of their member-ownership. But do locally rooted cooperatives act in a manner comparable to other types of local enterprise (Tolbert et al., 2002)? We know that in the United States some cooperative sectors are grouped with corporate models regarding state statutes, whereas cooperatives in the electric industry are typically left to be regu-lated by agents, market actors, states, and the membership served by the coopera-tives. How do these different governance arrangements influence the operational attributes and processes of cooperatives?

Significantly, little attention has been given to utility cooperatives, particularly in the energy sector, despite their penetration into vast swaths of the United States and their central importance to the livelihoods of over 42 million Americans. What if we were to structure substantial portions of vital industries as cooperative owned? While the United States may be held up as "the world's great exemplar of corporate capitalism," America is also home the world's largest cooperative business sector (Hansmann, 2000, p. 1). Electric cooperatives, in particular, comprise a significant share of the American electric utility industry (905 electric co-ops owning over half the electric grid. The cooperative electric power sector is no niche player. Considering the overlap between electric cooperative service territory and prime wind energy development resources, we can speculate that, if supported by consumer preference and optimal public policy, co-ops could play an even greater role in deploying next-generation wind and renewable energy. Whereas the literature has addressed many community aspects of investor-owned hydrocarbon energy development, it is sparse on the community outcomes of electric cooperative wind energy.

This book focuses on contextualizing the community economic development outcomes of wind energy development by both cooperative electric utilities and IOUs. While IOU wind energy has been touted for its community economic development capability, community economic development theory posits that community-based wind energy development could serve as an innovative, community-enhanced model of ownership for rural development. This potential has been mentioned in the community wind literature from organizations in Minnesota, Oregon, and Maine, in treatments by a few scholars, and by analysts at the U.S. Department of Agriculture (USDA). Most studies, however, focus primarily on municipal or collective arrangements of wind power ownership, as opposed to cooperative models. This book addresses these gaps while advancing new contributions to how policy makers, developers, planners, and researchers think about the intersection of government policy, institutional design, ownership, fit, and community economic development.

Conducting the Research: Methodology

The book is in part an exploration of the variance in how two models of ownership of the firm influence community development practices and outcomes. This exploration occurs through two case studies. The emphasis of this book is to understand better how the stakeholders (community, the firm, and extralocal actors) interact, what the products of these interactions are, and the applicability of the given ownership model to the resource system (electricity).

A comparative case study methodology is utilized to address the broader question, how does wind energy development interact with community economic development? Case studies are "chosen for the likelihood that they will offer theoretical insight" (Eisenhardt & Graebner, 2007, p. 27). Insight might come from an observed phenomenon or case (e.g. the rise and fall of a boomtown; Broadway & Stull, 2006), or the outcomes of a community development project (Zacharakis & Flora, 2005) might serve as the unit of analysis bounded by time and activity (Creswell, 2008, p. 15; Yin, 2009). The case study method, often used in public administration, sociology, and elsewhere, is a small-N, nonrandom, purposive empirical research strategy (Yin, 2009). The case study methodology used by scholars seeks to analyze complex system processes through the triangulation of the broad range of data collected and the incorporation of actor perspectives (Tellis, 1997). The findings herein should be considered exploratory.

The interest here is primarily in the structural attributes of the community, the wind energy utility, and the community outcomes of these interaction processes, as well as government and market actors influencing this system. These three units of analysis are assessed through two cases: the cooperative PrairieWinds wind farm in Ward County, North Dakota, and the IOU Twin Groves wind farm in McLean County, Illinois. The broad research question could then be rephrased from a more technical perspective: how does wind energy development interact with local social structures (the community action situation)?

The underlying hypothesis is that the community economic development outcomes vary depending on the ownership model, hence the subquestion, how does the ownership model of the wind energy firm interact with community economic development? Though a number of wind energy ownership models exist, this study only compares two: the investor-owned and the cooperative-owned utility. While there is an abundance of investor-owned wind farms in the United States, researchers have a much smaller population of cooperative-owned wind firms to choose from: at the outset of this research, only two utility-scale cooperative wind farms existed, both owned by Basin Electric Power Cooperative. While this study is like many case studies in using an intentional and often nonrepresentative sample, a great deal can be learned and knowledge advanced from the study of an outlier, providing differential findings in the subject matter.

A thorough analysis of the complex levels of social systems at play is critical for understanding the outcomes within the community (Miraftab, 2004, p. 92). The case study methodology lends itself to better understanding community power dynamics by discerning "socially structured and culturally patterned behavior of groups, and practices of institutions" (Lukes, 2004, p. 26). The methodology attempts to be thorough in "observation, reconstruction, and analysis" (Tellis, 1997, p. 3), incorporating the voices of actors embedded in the case. Here, the methodological approach uses the institutional analysis and development framework (IAD) to analyze the interactions between the wind energy firm and the community (Ostrom, 2005).

Data Collection and Analytic Methodology

Data sought and collected are predicated on an IAD-informed systems approach, meaning the community and the wind firm are not isolated but instead nested within a complex social structure, with each level having an impact on the interactive processes. The attributes of a given case, the contextual realities, and the participants are considered to varying degrees at multiple levels of analysis.

Three levels of analysis are utilized under IAD: (a) local or "operational level, the level of day-to-day decisions made by participants in any institutional setting" (i.e., the wind firm); (b) organizational governance or "collective-choice level, the level determining the operational activities and results through specific, domain-focused institutional and organization structures and operational rules" (i.e., the community), and (c) "constitutional-choice or policy level, the level that defines the broad parameters of social action and social order creating the foundations for the institutional arrangements and the rules to be used in crafting the set of collective-choice rules that in turn affect the set of operational rules" (Aligica & Boettke, 2009, p. 86).[5]

Understanding the contributions and costs of an electric utility requires attention to extralocal actors. The very design of the national electric grid overlays every municipality in the United States. These levels (external to the community/action situation) are analyzed through an assessment of government energy policy and relevant market actors. I then triangulate the analysis against the findings from the field to assess how these social forces influence aspects of the community economic development process. The community members can be represented as individuals, groups, or organizations. These characteristics are used to analyze the data and describe the motivations and level of influence that given actors have within the action situation.

From the perspective of the Bloomington School, an action situation broadly comprises seven attributes affecting the "actions of participants," though to a differing degree based on contextual realities (Poteete et al., 2010):

1. the set of participants confronting a collective-action problem,
2. the sets of positions or roles participants fill in the context of this situation,
3. the set of allowable actions for participants in each role or position,
4. the level of control that an individual or group has over an action,
5. the potential outcomes associated with each possible combinations of actions,
6. the amount of information available to actors, and
7. the costs and benefits associated with each possible action and outcome.[6]

Data collected from participants were analyzed based on these seven participant attributes in an attempt to reconstruct the action situation and weave a cogent narrative that addresses processes and patterns relevant to the research questions (more on that below).

Archival analyses and semistructured interviews either in person or via phone or e-mail contextualize the action situation (i.e., the interaction between the wind utility and the community). Data collection started in 2009 and stopped in 2012, to represent better the issues facing the wind energy firms and their communities during that specific period. An archival analysis was used to explore such relevant archival documentation as official organizational documentation (agency or business literature) and how local media outlets depicted wind energy development processes. This then set the tone for the fieldwork undertaken.

Fieldwork added access to on-the-ground realities that would have been missed offsite. Due to the relative scarcity of research endeavors focused on wind farms, this fieldwork, coupled with other sources of data collected for analysis, was of critical importance. Fieldwork furthers our knowledge of these social systems so that researchers might have access to more vibrant, more complete information with which to better inform policy (Poteete et al., 2010, p. 88).

A total of four weeks of fieldwork was carried out in the two communities (roughly two weeks in each community). Fieldwork in Ward County, North Dakota, occurred in August of 2011, and fieldwork in McLean County, Illinois, occurred in August–December 2011. Purposive access was sought to interact with national- and local-level actors through gatekeepers in academia (initially through university extension offices) and my own contacts in the wind and cooperative energy sectors. Fieldwork comprised observation of and interviews with participants in the action arena (reputational leaders, landowners, government officials, wind energy developers, and other wind-energy-related stakeholders that snowballed on an as-needed basis). The semistructured interviews were carried out with these individuals in an attempt to assess local power structures, community interaction processes with the wind energy firm, how governance functions, and how critical material and social resources are distributed locally.

The research then incorporates the perspectives of those who live within the host community to paint a more complete picture of the interactive processes and community outcomes (Poteete et al., 2010, p. 52). Eighty-six total interviews were conducted with forty-nine people. Among these, thirty-four interviews were conducted with fifteen national-level actors at cooperative business associations, the U.S. Department of Energy (DOE), and the USDA (the nationally oriented, investor-backed wind energy associations were not accessible for this research project despite many attempts). Thirty-two interviews were carried out in Illinois with twenty participants in the investor-owned wind energy sectors, agricultural interests, developers, landowners, and local public officials. Twenty interviews were conducted in North Dakota from a pool of fourteen participants in the cooperative, wind, and public sectors.[7]

The questions in the interview protocol were geared toward an assessment of the local community social structure (Zacharakis & Flora, 2005). The questions did not produce anticipated outcomes but instead elicited responses to identify locally oriented actors with direct involvement in the wind energy development process; the responses provided additional insights useful for further analyses. These data were used to assess the actors engaged in the localized wind energy development process. The interconnectedness and characteristics of the community-level actors are weighted against the patterns of structured interactions to assess better who most benefited from the structural and material returns.

The research design is concerned not only with the structure of the community but also with how actor networks interact with the wind farm and the community, in order to tease out any tensions or mutual benefits. Data collected from archival analyses, participant observations, interviews, and social network analyses are applied to identify the distinct community social structure and how it interacts with the wind energy utility.

Analyzing the Data

Taking a bird's eye view, the data analyses applied to this research can be summarized as follows:

1. perform a snapshot assessment of the polycentric attributes of the IOU and cooperative wind energy systems,
2. contextualize the host communities, and,
3. critically analyze the interactional patterns observed between the host communities and the wind farms.

IAD is used for analytic triangulation.[8] It is a consistently tested and utilized framework embracing the inherent complexity of institutions and social systems (Ostrom & Kiser, 2000; Poteete et al., 2010).[9] In this case, the study applies the community to the action situation within IAD.[10] The IAD is useful for understanding both the exogenous variables that interact within the community and the endogenous community interactions, specifically the interaction of wind energy development with the community.

IAD is employed to structure identified patterns of interactions, to understand how actors interact and how these collective interactions impact the community. Social structural impediments are of central concern to community economic development; therefore, understanding the local social structure is critical. Contextualizing the structure of the community helps assess whether the community and the wind energy actors have polycentric or monocentric tendencies and what that means for the livelihoods of people living there. Here interest acutely focuses on how the community interacts with the localized wind farm and whether that interaction lends itself to community development. Methodologically, data collected are used to discover the structure of the community interacting with the wind utility, with the partial assistance of a simple social network analysis.

Aligica and Tarko's (2012) logic structure of polycentric systems complements the structural analyses of the wind energy system. The logic structure accounts for the three basic features: (a) "the multiplicity of decision centers," (b) "the overarching system of rules," and (c) the "spontaneous" or purposive "order generated by evolutionary competition between the different decision centers' ideas, methods, and way of doing things" (p. 257). The diagnostic is used in this research to assess the structural features evident within the wind energy sector, and whether that structure tends to allow for institutional innovation (getting at the question of why only one co-op owns a wind farm). Because this is a case study and not a large-N analysis, the observations and findings provide a glimpse into these systems but cannot definitely answer the questions posed.

The paths available for alternative institutions within the rapidly growing wind energy sector are essential to understand, because absentee firms, or outsiders, have been linked to community atomization. Aligica and Tarko (2012, p. 254) note one potential explanation: "outsiders" are agents not subjected to the same system of rules as "insiders"; thus, the potential for disruption is evident. Certain institutions may be a better fit regarding community development, and social systems may help determine which institutions are applied. Therefore, an assessment of the centricity of the wind-energy industrial social system allows for enhanced capacity to analyze and assess the state of the community structure, the local dimensions of power (Brennan, 2009; Lukes, 2004), and how key institutional actors develop the community.

Finally, it is not enough to assess systems; one must also assess the organizational manifestation itself, the wind firm. The ODPs listed in Table 1.2 are utilized to enhance analyses of each type of wind energy firms.[11]

The ODPs serve as a diagnostic to assess organizational robustness or the ability of the firm to adapt to disturbances (Ostrom, 2005, p. 258). The ODPs have been applied to governance analyses of common-pool resources, as well as public and private goods (Aligica & Boettke, 2009). Using a valid analytic for assessing organizational robustness is critical for several reasons. Government-driven wind energy policy may privilege one type of institution over another. Perhaps a particular institution is more prone to poor owner governance or market shocks. Should that be the case, "we can conclude that there are ways of organizing governance that increase the opportunities for adaptation and learning in a changing and uncertain world with continuing advances in knowledge and technologies" (Ostrom, 2005, p. 257). In doing so, we can bring about greater stability in policy and development while enhancing the day-to-day lives of the people on the ground that intersect with these dynamics.

Concluding Thoughts on the Methodological Approach

Complex social dilemmas require sophisticated tools for analyses. Bloomington School analytics were designed for that exact reason. IAD allows for input-output analyses (the social forces that interact with the community), whereas the polycentric logical structure diagnostic allows for a refined evaluation of the social systemic attributes (the question of government policy and market influence). The ODPs then allow the analyst to understand better how the structured design of the firm may interact with the host community, contributing to enhanced knowledge of the institutional design of the wind energy firm.

Embedding these analytics within the case study methodology is a potent approach toward the exploration of heretofore little-understood social dilemmas. The emphasis on outliers or uniquely interesting social phenomena allows the analyst to observe noteworthy social processes that may be obscured in statistical studies.

CHAPTER TWO

Case Study—The Investor-Owned Wind Farm

Wind energy is the fastest growing segment of the electric energy generation market. The rapid growth is in part due to three factors. First, the uncertainty of proposed emissions regulations—as well as political pressure groups—has made it increasingly difficult to build new coal and nuclear power plants. Wind energy bypasses many of the debates associated with these exhaustible legacy energy sources. Second, incentives and subsidies at the federal and (some) state levels further propel wind energy development. The most advantageous subsidies are available only to those entities and investors with a tax appetite[1] to fully utilize the tax credits. Third, even absent subsidies, wind energy has become price competitive with legacy fuels such as coal and nuclear.

Federal incentives for renewable energy development in the United States have historically consisted of the investment tax credit and the production tax credit (PTC), as well as the accelerated depreciation benefit for renewable energy property: the Modified Accelerated Cost Recovery System (MACRS) and bonus depreciation. Both the investment tax credit and the PTC provide financial incentives for development of renewable energy projects in the form of tax credits that can be used to offset taxes paid on company profits. Given that many renewable energy companies are relatively nascent and small, their tax liability is often less than the value of the tax credits received; therefore, some project developers cannot immediately recoup the value of these tax credits directly. Typically, these developers have relied on third-party tax equity investors to monetize the value of the tax credits (Steinberg & Porro, 2012).

The tax-equity partners are by and large major financial firms, incentivized by government energy policy to invest in wind energy. A so-called tax-equity partner is usually a bank or other large company with a hefty tax bill that can take advantage of the tax credits and either provide a loan for the project or buy it. Bank of America, U.S. Bancorp, and other banks, as well as corporations like Google Inc., have used the tax-equity structure to invest in solar power and other renewable-energy projects (Tracy, 2012). Green energy is in a sense also big finance.

The tax-equity partnership allows a wind company to partner with a large firm with a significant tax liability. The tax liability is reduced—not eliminated—via tax credits (not grants or direct subsidies in the form of cash payouts) provided by either the PTC or investment tax credit. The federal government reduces potential tax revenue generated from new renewable energy production, incentivizing investment in and growth of wind energy by profit-driven firms who are then allowed to capture their revenue surplus for enhanced returns. Wind energy companies may profit further from this policy by monetizing the tax credits through an overhead fee for the privilege of partnering on the project (association official, Nov. 3, 2010).[2] In an era of energy market volatility, wind serves as a safe tax shelter and investment hedge for deep-pocketed investors from the financial sector. It should come as no surprise that 98 percent of all wind farms are structured as for-profit limited liability companies (LLCs; (AWEA, 2016).

My first case study examines McLean County, Illinois, host to three major wind energy farms, all investor owned. The case study analysis focuses on one of the wind energy firms, the investor-owned Horizon Wind Energy, a subsidiary of EDP Renewables.[3] As will be discussed, Horizon is noteworthy because it is the first wind energy developer in McLean and helped lay the foundation for future prospective wind energy developers to operate within the county. The purpose here is to analyze and assess the development implications of the investor-owned wind farm within its host community. It is therefore vital to understand the social forces (the local community, market drivers, and government policy pathways) that compelled Horizon to be the first wind energy developer in McLean and how the community at large responded to the prospects of a major wind energy project being located in their community. The chapter ends with some concluding thoughts on the policy, as well as the ramifications of investor ownership on the local community.

McLean County and the Investor-Owned Horizon Wind Energy LLC

McLean County in central Illinois shares many of the characteristics of other U.S. counties with metropolitan centers (Isserman, Feser, & Warren, 2009). It has a relatively prosperous urban-city core, Bloomington-Normal, surrounded by a rural region experiencing population flight and a diminishing tax base (local official, Sept. 5, 2011). The rural areas are in a state of relative decline as the urban core of Bloomington-Normal consolidates regional resources within its borders. Much of the decline has to do with the challenges facing rural communities in harnessing community economic development opportunities for enhanced local livelihoods. A local university extension official noted: "Outer lying areas feel left out of development. The area is becoming urban . . . diverse. It's where urban and country meet" (Oct. 24, 2011).

What makes McLean County interesting for this analysis is the relative economic

Table 2.1. McLean County, Illinois, comparative economic data, 2010–2014

Attribute	McLean County	Illinois
Population	173,166	12,859,995
Unemployment rate (June 2012)	7%	9%
Population: percent change, April 1, 2010 (estimates base), to July 1, 2015	2.10%	0.20%
Percent below poverty rate	14.30%	14.40%
Per capita income in past 12 months (in 2014 dollars), 2010–2014	$30,728	$30,019
Median household income (in 2014 dollars), 2010–2014	$61,995	$57,166

Source: http://quickfacts.census.gov/qfd/states/17/17113.html; http://research.stlouisfed.org/fred2/categories/150.

stability brought to the community by a diversified economy comprising a significant number of transnational firms. McLean's economic health relative to Illinois appears more robust (see Table 2.1), with access to many built and cultural assets. The local economic development council has been marketing McLean from this vantage point for decades: "Endowed with excellent location and transportation, advantages, diverse population and employment sectors, and enviable community assets, the area has become a sought-after site for commercial and industrial development" (Bloomington-Normal Economic Development Council, 2013). The development profiles listed on the economic development council's website go on to note McLean's centrality to six major roadways, four railroads, and a regional airport. Such attributes are attractive to large-scale wind energy firms seeking access the road, rail, and grid for development and operations.

Agriculture plays a significant role in the lives of McLean County residents. Community identity is interwoven with the surrounding flatlands and row crops. Agriculture associations provide a considerable amount of local-level employment. The influence of agriculture is reinforced by the presence of the headquarters of the Illinois Farm Bureau, which is quite active in local community life (staff are granted on-the-clock volunteer time). The local agricultural community is further distinguished as the highest corn- and soy-producing county in Illinois (Steever, 2011), with many prominent local farmers working in collaboration with the biotech giant Monsanto at their high-tech agricultural research station to advance new crops.

Agriculture's prominence is further secured by a county zoning ordinance, which inhibits the encroachment of urbanization and sprawl on the existing stock of farmland in the county. "In 1974 a zoning ordinance placed severe restrictions on residential zoning in the ag[ricultural] district. It's been a bit of a problem for some people in the city who want to live in the country. They buy up land undesirable for farming for the purposes of building their country home, only to find out they're prohibited from building out there" (local official, Dec. 13, 2010).

A local extension official, when asked about the effectiveness of the zoning ordinance, agreed that it preserves farmland but offered a caveat: "Bloomington-Normal is in Veterans Parkway. We have sprawl. We have malls. If you have a major project with a lot of money attached, the ordinance all of a sudden becomes more flexible" (Oct. 24, 2011).

Regardless of the general effectiveness of the ordinance in preserving the local farmland, the agricultural scene employs relatively few people compared to other local industrial sectors (roughly 2 percent during peak farm season; US Department of Labor, 2015) and has been steeply declining since the 1970s. In other words, the agricultural identity persists despite its limited distributed economic impact.

Economic growth seems to be a driving force in McLean County. Much of day-to-day life and economic activity happens in McLean's twin cities of Bloomington and Normal, where the business community plays a significant leadership role. The chamber of commerce is perceived by some participants as the go-to group for catalyzing any local development initiative (the chamber founded the Bloomington-Normal Economic Development Council).

The "chamber crowd" comprises sizable numbers of corporate partners and established local business people (extension official, Oct. 24, 2011). The headquarters of State Farm Insurance—itself a policyholder-owned mutual—is located in Bloomington-Normal, employing almost 16,000 locals as of 2011. Transnational corporations such as Country Financial, Mitsubishi Motors, Bridgestone/Firestone, and Verizon collectively employ over five thousand workers from within the twin-city boundaries. The twin cities also serve as an area hub for medical treatment at its hospitals (local official, Sept. 5, 2011), which employ almost three thousand, providing specialized medical care to individuals throughout the region.

While locally produced commodity-agriculture may have limited community-wide economic impacts, interstate agribusiness interests help bolster the area's affinity with agriculture. The Illinois Farm Bureau's headquarters is in the area, as is the Illinois Agricultural Association and the ag-supply provider Growmark, Inc., one of the nation's largest ag-supply cooperatives.

McLean's economy is further assisted by government investment in human capital and public services for the county's 172,281 residents (U.S. Census Bureau, 2015). The University of Illinois, Illinois State University, Illinois Wesleyan University, and Heartland Community College are major drivers in cultivating the human capital necessary for the community to compete in a global marketplace (over 25,000 students were enrolled during the 2011–2012 academic year). Combined, these entities employ over six thousand high-skilled workers within McLean and ensure base human capital needs and persistent workforce regeneration.

A thriving food service sector reinforces the professional workforce. Residents emphasized McLean's place as the "Restaurant Capital of the World" based on per capita saturation of restaurants locally (though numbers were not provided to back the assertion) (local resident, Sept. 6, 2011).

The locally rooted industry seems to be able to handle Bloomington-Normal's population growth rate well, which outpaces the Illinois state rate, providing jobs that pay more than the Illinois median (see Table 2.1). These organizations offer a diversified mix of quality jobs for the region through their involvement and positionality in the global marketplace, stable tax income, and an engaged workforce involved in local civic life (see Table 2.2). The corporate culture in McLean encourages and incentivizes professional development through involvement in local charities, partaking in nonprofit boards, and running for local-level political office (extension official, Oct. 24, 2011; association official, Nov. 11, 2011).

It comes as no surprise that, with Bloomington-Normal's mix of corporate and state investment, the surrounding county boasts a stable economy with good jobs. But that stability is not based on local robustness. McLean's prosperity is increasingly embedded in and dependent on global markets. Even the dark, fertile farmland, often mentioned as a core identity of the community, is embedded in global commodity markets (extension official, Oct. 24, 2011; local resident, Sept. 6, 2011). A local university official stated: "We are proud of having some of the world's most fertile farmland. We rank toward the top in the world in productivity per acre" (Sept. 2011). The area's row-crop farmers in general bypass local markets, taking their commodities to the global marketplace (University of Illinois Extension, 2013), which is peculiar considering the robust local food service sector and persistent demand for local food.

Recently, younger, nontraditional farmers and members of the professional business community have taken steps to start a new grocery cooperative, as well as a local foods hub. This development is politely challenging the community's deep dependency on global markets. A regionally sourced dairy farmer noted: "Local foods are a niche. A nice niche, don't get me wrong. But corn and soy certainly dominate. I would say 99.9999999 percent of the ag[ricultural]-base here is corn and soy" (Dec. 1, 2011).

Interestingly, many of the core supporters come from the existing transnational agribusiness community. Generally speaking, the farmland itself is treated not as a source of local subsistence to guard against larger social system shocks but as

Table 2.2. Major McLean County employers

Company	No. employed
State Farm Insurance Companies	14,528
Illinois State University	3,275
Country Financial	2,049
Unit 5 Schools	1,754
Mitsubishi Motor Manufacturing	1,270
Advocate BroMenn Medical Center	1,157
Growmark, Inc.	932

Source: http://www.bnbiz.org/uploads/documents/uploader/470_uploader_atanequry3.pdf.

something geared toward financial outcomes via exporting product to global commodities markets.

Globalization has worked for many of the residents of McLean for decades. The wind farms dotting the farmland and lining the highways continue this tradition of global markets rooting themselves in the fertile midwestern soil. Horizon Wind Energy is itself part of EDP, a corporation headquartered in Madrid. The local wind farms represent not a cultural shift as much as a continuation of long-established local economic development policy linked to global economic processes.

Despite the absentee ownership of Horizon, there is a great deal of local boosterism for wind energy (Miller, 2007a). The wind energy companies continue to fuel the boosterism (Ford, 2006a) by tapping into local desires for increased economic development. From the vantage point of local policy makers, absentee ownership is not viewed as exploitive or even extractive but as a contribution to prosperity via utilization of cropland to produce electricity (instead of food), jobs, lease payments, and an enhanced tax base.

In other words, the numerous wind farms play a part in the social life of McLean County, but not a wildly influential role. The wind farm is but one set of social processes playing out in a much larger arena, rife with numerous interactive social processes. Wind energy production is an extractive process in that the value of localized natural resources is maximized for consumption and profit elsewhere (though wind energy extraction is subtler and less agitating, with a cleaner facade than industries such as coal). Wind is rarely viewed as a resource in and of itself, meaning contestation over harnessing the energy from wind would be a relatively new phenomenon. The argument exists that there is a level of exploitation, but the exploitation is shrouded in subtle complexity ("wind theft" would be a difficult accusation to prove, for obvious reasons). As long as the wind energy company can align its perceived values orientation with the host community, individual landowners, and political leadership, there isn't a great deal a wind farm will need to do to interact with the local community once it is built and operational. The turbines become just another part of the local landscape, quietly generating revenue for the investor-ownership, landowners, and taxing bodies. And the small, locally based Horizon workforce plays a role in enabling the continued development of wind energy in McLean. The wind farm doesn't appear to be extracting anything from the community; wind farms like Horizon position themselves as net community builders, converting a previously unused resource into something productive.

While McLean has relative prosperity, with a highly educated and trained workforce, those factors are not what primarily attracted Horizon to the area—it is not the driving force behind wind energy development. A mix of plentiful wind, abundant and open land, easy access to the transmission lines, and consumer demand backed by government subsidy helped pave the way. Arguably the only prerequisite that a wind-resource-rich area like McLean needed to meet was to be welcoming, with a business climate speaking the language of tax incentives and subsidies.

Building and operating a wind farm is a long, drawn-out process for the uninitiated, and is ever changing for the seasoned wind energy development professional. The electric energy sector is a complex amalgamation of capital-intensive investor-owned, municipal, and cooperative businesses interacting through tight regulatory bodies, necessitating the use of legal experts, engineers, and politically influential individuals. But the development outcomes of wind energy are subtler than one might presume with such a visual spectacle requiring massive coordination efforts and infusion of upfront risk capital. The devil is in the details of the wind energy development process.

A New Neighbor?
How the Transnational Horizon
Wind Farms Came to Be

The case here focuses on Horizon alone. The goal is to analyze one specific organization, its business value chain, and its total effect on community economic development. Plus, Horizon is a major player in this particular case. It was first to the area in 2001, the first to break ground and become commercially operational in 2007, and the first working with local administrators to clarify zoning ordinances, clearing the way to make additional local wind energy development more accessible and attractive to other developers. Horizon is the company that helped set the bar locally.

Horizon has been ambitious in McLean. Two of Horizon's twenty-seven U.S. wind farms are in the county, comprising 240 turbines; moreover, Horizon has proposed hundreds more and another 200 turbines in surrounding counties (Ford, 2010). However, Horizon isn't the only player in the region. Oregon-based Iberdrola Renewables has explored the development of 337 turbines in the same service area. There have also been significant proposals put forward by Navitas Energy (143 turbines) and Invenergy LLC (100 turbines) (Shults, 2006). A local zoning official believes it is feasible that 400 turbines will go online in the next couple of years. Wind energy development has the potential to change the landscape and culture and to dramatically enhance the tax revenue base.

Nationally, there is a rush of sorts to occupy areas where the development processes have been formally incorporated into local public policy (Sapochetti, 2010). If the sociopolitical mood is amenable, the land, wind, and grid are finite; speculators must strike while the iron's hot before the competition gets there first.

McLean was attractive to Horizon for many reasons. Initially, the county met many of the biophysical prerequisites necessary for speculative development to begin. (a) The average wind speed in the vicinity met the calculations to make a wind farm viable (though as turbines grow increasingly more efficient, wind energy developers will be able to utilize resources in regions with lower wind speeds, making more communities potential prospects for wind energy development; Loomis & Carter, 2011, p. 12). (b) Existing transmission lines had excess capacity. The costs of developing a wind farm are expensive enough without the added burden of building new transmission capacity. It is ideal to find a wind regime as close as possible to transmission capacity to limit infrastructural investment needs. Illinois transmission was designed for growth. Fortunately for wind energy developers, the Illinois electric grid was designed primarily to send power from the rural countryside to be consumed by the Chicago electric market (association official, Nov. 3, 2010). Transmission infrastructure spiders out from the Chicago area southward, intersecting with wide-open spaces and blustery wind. McLean happens to be one such area in which this intersection of transmission and wind regime occurs. (c) Vast tracts of relatively contiguous plots of land were available. The closer the turbines are to the transmission infrastructure, and to one another, situated on vast tracts of empty parcels of land, the lower the connectivity overhead costs.

Beyond the biophysical prerequisites, because electricity generated locally is transmitted to the national grid, numerous sociopolitical processes play into wind energy development, at varying scales.

McLean County Is Open for Business!
Enabling Wind Energy Development

Wind energy speculators do not have to guess as to where wind energy resources may be in abundance. A broad array of wind data has been publicly accessible through the DOE National Renewable Energy Laboratory, where the wind data estimates are updated on a regular basis. McLean and many other counties are in prospective wind energy developers' databases, waiting to be assessed for development.

While the National Renewable Energy Laboratory's wind energy data sets are useful for wind energy speculation, developers must still collect on-the-ground data to assess operational feasibility. Wind energy developers undertake localized precision testing to isolate ideal conditions for wind farm infrastructure. Some areas support independent programs for "wind prospectors" that collect this on-the-ground data to be used for wind energy development interests.[4] The data collected are then housed on publicly available websites and used by these prospectors to entice wind energy developers. According to an official in one of these programs, "these projects have caught the attention of some companies. The Rail Splitter wind farm on your way into town came about because a prospector collected the data, and worked with us to talk to a wind company. See, they demonstrate not only feasibility, thereby lessening the wind company's development overhead, but also a welcome feeling by the community" (university official, May 10, 2010).

The testing assesses if the land is capable of hosting a large wind farm. If yes, developers must make sure the wind speeds and patterns align with the available land through precise measurements. Testing is done through the utilization of a meteorological (MET) tower over a one- to two-year period to better isolate the ideal placement of the wind farm for a reliably productive wind flow at the local level.

Often, before the wind developers spend the time to deploy a MET tower or hire a contractor to assess overall transmission grid accessibility, they determine community sentiment. A feasibility prerequisite is a determination of the community's willingness to host a wind farm; if outright, mobilized hostility is apparent, the wind developers are likely to look for a more welcoming community elsewhere.

Should initial contact with the community look positive, a two-tiered process begins locally, involving land development and community cultivation. The wind energy developers may begin the speculative land development process if the assessed biophysical requirements and investment projections look favorable. Wind developers start the long process of cultivating the community (opinion leaders, policy makers, and landowners) by preparing the population to understand what hosting a wind farm entails, as well as the material benefits they can expect.

Such a capital-intensive project requires significant investments of human re-sources and time. "Horizon had years of broad community conversations before they committed to the wind farm (they were clear they were noncommittal)" (lo-cal official, Dec. 13, 2010). Horizon began a public dialogue with the communities within McLean for wind energy development in 2001. The wind farm achieved commercial operation in June 2007.

As is typical, a multipronged, parallel process began in which Horizon develop-ers started talking to two groups serving as critical determinants of development: local government officials and landowners. Any major wind farm will need access to land, most often owned by a collection of local landowners. The development staff must know first that a group of landowners are open to signing long-term leases and that the local government bodies are willing to issue new building and zoning ordinances to accommodate the unique attributes of a wind farm. According to a zoning official: "We worked hard to make this happen. We had to learn a great deal about what wind energy projects would do in the County. But now we get a lot of calls from other communities being approached to build out wind farms. It was a long ordeal, but we have become experts in the field" (Dec. 13, 2010).

Horizon hired two individuals from within the local community to serve as its lead project developers. They were seen as young, ambitious, and knowledgeable of local mores. The individuals chosen to be local-level developers must by the very nature of the job be highly entrepreneurial and able to adapt to sudden changes on the ground. These professionals are project dedicated, expected to devote two to five years to see a given project to fruition.

Despite the significant amount of capital wind energy development and the preci-sion engineering studies, wind energy developers are still developing best practices. When asked about the extent to which developers shared best practices among other developers, one of Horizon's development staff responded: "Well, all devel-opers share information with corporate, and corporate then shares it with us. We follow trade publications, the news, and we even apprentice new developers from other wind energy companies" (Nov. 9, 2011). When asked about the best practices used for community organizing, one of the local wind developer's response was, "Do you have a development handbook? I would love to have one of those!" (Nov. 8, 2011).

It seems as though there is not yet an official how-to guide for individual wind energy developers. But the interactive, iterative processes crafted by staff working across firms signal that some best practices are arising. Successfully developed com-munities like McLean serve as valuable laboratories, enhancing the development capabilities of wind energy firms.

The staff viewed themselves more as economic development professionals seeking broad-based buy-in as opposed to community organizers advocating for social change and justice. It was interesting that in conversations with the development staff specific community organizing strategies or techniques were never mentioned, even though many of the mechanisms utilized were reminiscent of traditional mobilization techniques. The Horizon developers set up community dinners with prospective local landowners, who were by and large from the farming community. Horizon asked committed landowners to host neighbor open houses in an attempt to trade on brokered social capital. Horizon also hosted open houses for the community at local schools and firehouses (Ford, 2010) in which staff handed out packets of literature and gave PowerPoint presentations. The use of respected individuals and civic institutions for meeting locals proved to be a reliable approach to harness social capital toward creating buy-in and engendering trust.

Horizon sees transparent trust building with the local community as essential. A Horizon wind developer stated: "Transparency is crucial. Everything has to be above bar. Perception is everything. No meetings are performed on short notice. All of our meetings, we give plenty of notice. We make sure to note that everyone gets the same amount for their leases (we have to avoid any inference of favoritism), and if in the future we build another wind farm with a more favorable lease agreement, then existing landowners will also get that" (Nov. 8, 2011). A second Horizon developer noted: "The more community outreach, the more you reduce lawsuits. Our practice is at least four public events" (Nov. 9, 2011).

Community interest appears to have been substantial from the outset, with the turnout for initial public forums reaching over three hundred participants. These initial community meetings typically covered the number of turbines, the impact on day-to-day life, and the new financial benefits for the community. The numbers quickly tapered off as it became apparent that only a handful of locals would be directly receiving a material benefit (local official, Dec. 8, 2011). Indeed, interviews with key leaders involved in wind energy development demonstrated that the breadth of stakeholders included in the standard wind energy development project is relatively narrow.[5]

As the wind farm went from vision to reality, the wind energy developers were typically tasked with providing numerous reports not only to meet regulatory requirements but also to appease local critics and mitigate the strength of organized opposition. In this manner, Horizon built ever-greater trust in that it demonstrated concern for many local issues by spending a great deal of time and money on these reports.

As the project advances beyond prospecting into actual development, the developers seek contractual options with landowners to guarantee that once build-out begins the wind firm has the approval to build on desired parcels of land. From Horizon's website:

> Horizon works with landowners who are interested in long-term business re-
> lationships. There are two phases to these relationships: the option phase and
> the lease phase. During the option phase, Horizon has the flexibility to execute
> the lease once the project is ready to be constructed. The option phase usually
> lasts approximately five to seven years and gives Horizon the time needed to
> measure the wind, secure access to transmission and obtain permits. In some
> cases, an extended option phase is necessary due to regulatory variations at the
> state or federal level or slower growth of the wind energy market in the region.

While developers from Horizon would not divulge the terms of an option, other wind developers claimed that many of these options are for five to ten years, providing annual retention stipends of $1 to $10 a year per proposed turbine (wind developer, Mar. 25, 2010).

Horizon did eventually exercise its options in 2006 and locked in a lease rate of $10,000 per turbine per year during the life of the contract. Neighbors directly adjacent to a parcel of land hosting a turbine were given the option by Horizon to receive "good neighbor agreements," a $1,500 annual payment. The payment is to compensate landowners for the burden of being near a turbine and serves as a waiver of legal liabilities related to shadow flickers from the rotating blades, noise pollution from the turbines, and other common complaints. Once Horizon locked in the necessary options, it was then able to move into procuring the capital to purchase the infrastructure.

Public Dollars for Private Gain: Mandates, Subsidies, and Market Demand

There must be demand for wind energy to encourage its development in a market economy. Typically, such market demand for wind energy comes from one of two sources. First, there is the consumer demand, usually stemming from environmental or price-oriented consumer preferences.[6] Broad consumer demand for renewable energy is substantive. Recent cost reductions in wind power and price hikes in carbon-based power have made it more attractive as well: "Three-dollar gas contributes to more wind energy. Political and policy issues also provide incentive. Energy is expensive, and it's not going to get cheaper" (Anderson, 2006). But consumer preference alone is considered a weak indicator in the electric energy sector. Wind may not be cheap relative to other sources of energy, particularly during the early stages of the industry (regulatory official, Oct. 25, 2010). Wind energy generation must be cost competitive to be effectively marketable. The same consumer who agitates for wind energy may become less demanding should wind energy prices increase and competing energy sources become cheaper, such as happened with the recent supply glut of natural gas. The fleeting desire of consumers to be environmentally friendly is not enough to justify significant capital investment in wind energy if prices aren't competitive.

Another way to drive up demand for wind energy is government regulation (AWEA, 2012a). Currently, the regulatory environment for coal and nuclear generation has made investments in more traditional energy sources unattractive (e.g., Illinois has a state ban on new nuclear development), which then pushes electricity generators to look for alternative sources to meet consumer demand.

Renewable portfolio standards (RPSs) provide another regulatory instrument to drive consumer demand. A mandated standard "requires electric utilities and other retail electric providers to supply a specified minimum amount of customer load with electricity from eligible renewable energy sources. The goal of an RPS is to stimulate market and technology development so that, ultimately, renewable energy will be economically competitive with conventional forms of electric power" (U.S. Environmental Protection Agency, 2009, n.p.). Such a policy then promotes one source of energy over another, critical for transitioning an industry dependent on finite resources contributing to climate change. Illinois is one of thirty-one U.S. states and territories with an RPS. The Illinois Power Agency Act of 2007 created an RPS of 25 percent, 75 percent of which must come explicitly from wind (Loomis & Carter, 2011, p. 6). Energy distributors are responsible for meeting the RPS, which they can do from purchasing wind energy from within or outside the state border.

Wind developers were keen to point out that RPSs have the positive externality of mitigating market volatility when wind energy is part of the mix (association official, Nov. 3, 2010). "Wind is an inexhaustible energy source, and it is free from fuel price volatility, which can contribute to the nation's energy security. Because of fuel price uncertainty, electricity supply portfolios need to be diversified. Wind power can help diversify electricity supply portfolios, which can then lead to relatively more stable energy prices, which benefits ratepayers in the long run" (Loomis & Carter, 2011, p. 10).

The volatility of the energy markets combined with enduring economic crises has made investors skittish about investing in capital-intensive energy projects left to strictly market forces. Most new wind-energy-generating facilities are addressing concerns about the volatility of the market and injecting stability by securing power-purchase agreements (PPAs)—a contractual agreement between a power producer and a power purchaser—before build-out begins, which is increasingly becoming the standard practice of wind and solar energy companies. This practice then reduces risks for businesses investing in wind energy generation. A state association spokesperson for the wind energy industry noted: "A PPA is a great hedge for investors in a volatile energy market. It used to be that coal was the cheapest energy source, and now it's natural gas thanks to new fracking technologies. But both of those energy sources will grow volatile again. A PPA locks in a guaranteed source of revenue for ten to twenty years" (Nov. 3, 2010). Wind energy becomes a hedge for investors wanting a stable return on investment and utilities wanting to project costs into the distant future. (Horizon was financed through an investment by the commodities division of Goldman Sachs, and the energy utility FirstEnergy procured a PPA for the energy generated by the wind farm.)

The federal government has created many incentives that encourage further growth in the wind energy industry by complementing cost competitiveness, achieved primarily through the MACRS and PTC. The MACRS is a subsidy in that it allows the wind firms to accelerate depreciation, allowing wind companies to write off more from their tax burden sooner than other types of energy generators and business: "Qualifying components of a wind farm are eligible for greatly accelerated depreciation deductions, typically over a five-year period based on the double declining balance method of depreciation" (Stoel Rives, LLP 2010, p. 78). The MACRS is one manner by which wind companies can retain their taxed earnings to either pass on cost savings to the consumer, provide a higher return on investment to the financiers, or invest in infrastructure.

The most important tool at the disposal of wind energy developers is the PTC, intended to drive the purchase cost of wind energy down through a tax credit—or tax reduction (association officials, Nov. 3, 2010, and Aug. 9, 2011). "The federal renewable energy production tax credit is an inflation-adjusted per-kWh credit that is applied to the output of a qualifying facility during the first ten years of operation" (Loomis & Carter, 2011, p. 9).

The tax credit (which is $.022 per kilowatt hour) serves two purposes. First, it lowers the cost of energy to the wholesale purchaser, who can then sell to the end consumer at a cost-competitive rate. Second, it creates an incentive for investors to infuse a project with capital; if the investor—Goldman Sachs, in the case of Horizon—has a significant tax liability, the tax credit reduces the wind energy ownership's tax debt.

The rapid rate of development into new locales is heavily dependent on the continuation of the PTC federal subsidy. Wind energy faces persistent boom-and-bust, up and down cycles when developers cannot rely on consistent public policy to prop up the growth of the sector. Communities dependent on manufacturing facilities and building tradespeople who install the infrastructure will be hardest hit in the years in which the subsidy lapses. This potential to effect community economic development outcomes certainly provides political leverage to the wind energy industry to harness for advantageous public policy. Indeed, the wind energy industry enjoys the support of red and blue state politicians alike, who view the sector as a net plus for job creation.

Subsidies, regulations, and economic incentives do not merely manifest at the national level. Local-level wind energy regulations and incentives also contribute to an environment that promotes or discourages wind energy development. Illinois provides two programs utilized by wind energy developers to reduce developer costs and encourage investment by wind energy firms in the state: the Illinois Enterprise Zone, and the High Impact Business designation. The Illinois Department of Commerce and Economic Opportunity (2016a, n.p.) describes the enterprise zone as follows: "The Illinois Enterprise Zone Act was signed into law December 7, 1982. The purpose of the Act is to stimulate economic growth and neighborhood revitalization in economically depressed areas of the state. Businesses located (or those that choose to locate) in a designated enterprise zone can become eligible to obtain special state and local tax incentives, regulatory relief, and improved governmental services, thus providing an economic stimulus to an area that would otherwise be neglected."

Businesses located or expanding in an Illinois Enterprise Zone may be eligible for an exemption on the retailers' occupation tax paid on building materials, an investment tax credit of 0.5 percent of qualified property, and an enterprise zone jobs tax credit for each job created in the zone for which a certified dislocated worker or economically disadvantaged individual is hired. Additional exemptions, such as an expanded state sales tax exemption on purchases of personal property used or consumed in the manufacturing process or in the operation of a pollution control facility, an exemption on the state utility tax for electricity and natural gas, and an exemption on the Illinois Commerce Commission's administrative charge and telecommunication excise tax, are available for companies that make the minimum statutory investment that either creates or retains the necessary number of jobs (Illinois Department of Commerce and Economic Opportunity, 2016b).

Horizon lobbied the Illinois government and received an enterprise zone exemption. Horizon used this designation to waive its sales tax obligation, which incentivized the firm to purchase a significant share of its building materials from within the state. Proponents argued that such a capital-intensive project would create many economic opportunities for local merchants. For example, McLean's wind farm sector has been a major boom for a gravel pit located in nearby Pontiac, which provided hundreds of tons of gravel for the concrete foundations for the turbines (L. K. Dodds, Feb. 18, 2011; Loomis & Carter, 2011).

Horizon qualified for the enterprise zone designation even though the location of the wind farms is in rural agricultural areas that were not economically depressed. This is a typical application of the enterprise zone that benefits wind energy developers, which speaks to the growing clout of the industry (regulatory official, Oct. 25, 2010).

More recently, Illinois has added another stackable tax credit/exemption with its High Impact Business designation:

The HIB program is designed to encourage large-scale economic development activities, by providing tax incentives (similar to those offered within an enterprise zone) to companies that propose to make a substantial capital investment in operations and will create or retain above average number of jobs. Businesses may qualify for: investment tax credits, a state sales tax exemption on building materials, an exemption from state sales tax on utilities, a state sales tax exemption on purchases of personal property used or consumed in the manufacturing process or the operation of a pollution control facility. The project must involve a minimum of $12 million investment causing the creation of 500 full-time jobs or an investment of $30 million causing the retention of 1500 full-time jobs. The investment must take place at a designated location in Illinois outside of an Enterprise Zone. The program has been expanded to include qualified new electric generating facility, production operations at a new coal mine or, a new or upgraded transmission facility that supports the creation of 150 Illinois coal-mining jobs, or a newly constructed gasification facility as a "Coal/Energy High Impact Businesses.

A qualifying High Impact Business may be eligible to receive the following: sales tax exemption on building materials, an investment tax credit, an exemption from state gas and electric tax, and a state sales tax exemption on personal property used or consumed in the manufacturing process or in the operation of a pollution control facility (North Dakota Industrial Commission, 2016, n.p.).

Horizon's first two wind farms did not qualify for this program since construction began in 2007 (future development projects may be eligible):

In 2009, the program was further expanded to include wind energy facilities. The designation as a Wind Energy/High Impact Business is contingent on the business constructing a new electric generation facility or expanding an existing wind power facility. "New wind power facility" means a newly constructed electric generation facility, or a newly constructed expansion of an existing electric generation facility, placed in service on or after July 1, 2009, that generates electricity using wind energy devices. "New wind energy device" means any device, with a nameplate capacity of at least 0.5 megawatts, that is used in the process of converting kinetic energy from the wind to generate electricity.

Beyond the more formal subsidies, Horizon received an additional grant from the state of Illinois for $2.2 million after the project had been finalized and financed (Coulter, 2006). The grant was more of a gift from the state than an incentive grant.

Illinois-based wind energy developers get access to significant federal, state, and local subsidies. Taken together, these incentives

- create market demand for wind through mandates as well as market demand,
- artificially reduce costs to the wholesale purchaser,
- drive down the end cost to consumers,
- reduce the tax burden of the wind farm and its investors, thereby increasing margins,
- increase the return on investment,
- entice massive infusions of capital from major firms,
- create a hedge against volatile energy markets while creating a sure-bet investment, and
- increase the growth rate of the wind energy industry and its economic impact.

There is a great deal of opposition to wind energy subsidy, often coming from established oil and coal lobbies. The narrative focuses on the privileging of wind and other renewables above and beyond the "establishment": "Tea Party groups and others, including The Heartland Institute and Americans for Prosperity, note the tax credits exist only to favor renewable energy over more traditional sources of power generation and are thus an obvious government manipulation of the market" (Glans, 2012).

If the argument is about subsidizing an energy source that is more expensive than other existing options, then yes, wind energy at spot market rates during the time of this study would usually cost more than other forms of energy generation, particularly coal and natural gas (though PPAs help to control volatility and costs, making wholesale prices more competitive, predictable, and attractive). This higher cost is not necessarily because wind is inherently more expensive, nor is it necessarily privileged relative to "establishment" energy regimes. One must consider the decades of subsidy that went into building legacy electrical generators such as coal and nuclear (and nuclear's government-provided insurance), not to mention the subsidization of rail transportation infrastructure, which externalizes a portion of the cost of coal distribution (Carson, 2010)—coal companies account for almost half of all freight rail traffic in the United States (Stagl, 2012). Also, there are various externalities related to coal (health concerns from air and water pollution, and environmental degradation from mountaintop removal and ash-waste ponds) that are rarely if ever factored into the real cost of coal. Opponents of the establishment energy regime claim that federal subsidy estimates add up to over $50 billion annually, eclipsing the annual subsidy allocated to wind, solar, and other renewables (Oil Change International, 2013). It seems as though the organized opposition to renewables has more to do with protecting legacy energy industries from new competitors than fiscally responsible, market-oriented public energy policy.

There are some legitimate concerns about how wind energy is subsidized. Wind energy development is pitched to a community as a net benefit. However, many of the local subsidies (enterprise zones, MACRS) are aimed at reducing the wind farm's tax liability. These profit maximization techniques are in direct conflict with the wind developer's claims of net community benefit; if given the opportunity, the developer will trim as much of its overhead as possible. Accelerated depreciation reduces the tax burden at just the time that the wind farm is paying down its debts and thus has a greater capacity to benefit a community financially, but contracts with communities typically do not address profit sharing, so the potential is lost. Plus, subsidies such as the enterprise zone, while marginal to the developer, are another missed opportunity for the local governments to benefit from wind farm tax revenue. In the case of Horizon, these two subsidies in Illinois seem to have done little to incentivize wind farm development and instead allowed Horizon's financiers to reduce the total expenditures that many other businesses are tasked with paying to local taxing bodies. While this may not be the case in other wind energy development scenarios, one can observe that issues of justice, fairness, equity, and voice indeed exist and promise to add to tensions with other local taxpayers who feel left out.

Accessing the Markets, Plugging into the Grid

A wind energy firm will need to access the electric grid, the gateway to the electricity market. Grid accessibility runs parallel to other relevant socioeconomic processes (securing capital, seeking tax benefits, and community organizing). The grid is itself a national construct, a common resource regime financed by a mix of private and public dollars, governed by a complex array of property arrangements. This commons requires monitoring of quantity of electricity transmitted from many actors since there is no way to isolate where an end user's power originated from—one may think of it like water: once it's pooled together in a reservoir, the end user cannot request a specific point of origin.[7]

All new energy generators in the region must establish a grid node by lining up a queue position through the Midwest Independent Systems Operators (MISO) to connect to the electric grid. MISO is one of the nation's independent system operators (ISOs) that regulates access to a designated regional grid. MISO serves as the gatekeeper, a monitor of the grid, and an arbiter of disputes. The remaining regulatory responsibilities, like retail sales, are left to state regulatory utility commissions, unless it is interstate, in which case it falls under North American Electric Reliability Corporation (NERC). MISO is a creature of the Federal Energy Regulatory Commission (FERC). FERC is a significant regulator of the grid, moderating interstate commerce of electricity. "The Federal Power Commission (FPC) was created in 1920 under the Federal Water Power Act for the purpose of regulating construction and operation of nonfederal hydroelectric projects. In 1977, when the U.S. Department of Energy was created, the FPC became the Federal Energy Regulation Commission" (Oil Change International, 2013). The DOE describes the system as follows:

> There are many individuals involved in running the grid. There are generator operators and transmission owners. But from a system perspective, one of the most critical entities is the independent system operator or regional transmission organizations. . . . They monitor system loads and voltage profiles; operate transmission facilities and direct generation; define operating limits and develop contingency plans; implement emergency procedures.
>
> Reliability coordinators also play an essential role. For instance, NERC . . . develops and enforces reliability standards; monitors the bulk power system; assesses future adequacy; audits owners, operators, and users for preparedness; and educates and trains industry personnel.

Horizon applied for access to the transmission line from MISO (MISO takes these queue requests on a first-come, first-served basis). ComEd (Commonwealth Edison Company) owns the transmission line most accessible to Horizon in McLean. Legally ComEd is not in the business of approving or denying access to its transmission line; MISO plays the role of referee. ComEd's transmission subsidiary collects rents in a nondiscriminatory manner on the usage and maintenance of the available capacity. By opening a transmission access request, Horizon agrees to enter into a number of engineering studies to make sure the grid can handle the increased capacity. The studies divulge the capacity of the transmission line at that node and any upgrade needed for connectivity. Once an interconnection agreement is finalized, Horizon is assessed a transmission access or "wheeling" fee (the cost for ComEd to maintain the line and for MISO to regulate the energy coming from Horizon over the grid).[8]

Local Buy-in, National Roadblocks

To finalize the project, Horizon went through the formal permitting process with McLean County's Department of Building and Zoning, and its board secured a

PPA with FirstEnergy (*Pantagraph* 2008) and then purchased the infrastructure to begin the build-out (Ford, 2006b). Horizon, like every other investor-owned wind firm prospecting for ideal development, had all of the requisites necessary to chase investment capital. By 2006, Horizon had made substantial investments and progress. The company spent millions, the community tentatively organized, and national, regional, and state actors approved of the wind energy generator's access to the grid.

Illinois law requires public hearings four to six months before issuing zoning permits. The local zoning board oversees a state-mandated, formal process in which it hosts public hearings, takes public testimony, and forwards recommendations to the McLean County board. The board then has end authority to approve or deny zoning permits (Local Official, Dec. 13, 2010). The venues chosen for public hearings are geared toward ease of access and to meet attendance estimates. The McLean County Zoning Board of Appeals held twenty public forums at ten locations (e.g., the Bloomington Center for the Performing Arts).

During such public hearings, numerous issues arise. The zoning board may take public comments and convert them into a prerequisite (i.e., defining turbine setbacks away from key infrastructure) to secure the zoning board's approval. But the developers also use these hearings to assess how to build community buy-in.

The wind developers at Horizon go into a community with a handful of company-approved "gifts" they may contribute to the host community if prompted to allay concerns and gain broad-based popular support. One Horizon developer noted they were approved to offer a fund in which they share a portion of Horizon's revenues with the community for economic development purposes (Nov. 8, 2011). In McLean's case, Horizon offered to create an escrow account to cover the costs of decommissioning the wind farm in the distant future but did not provide for a community development fund since such a request was not made.

Horizon had sought building permits extending a total of five years beyond McLean's standard two-year window. This extension is to account for hiccups along the way (cost increases, access to capital supply, expiration or extension of federal and state subsidies, and time to make a case with the local landowners and community members) (Ford, 2010). Horizon was wise to make such a move as many problems arose before and during the construction process (Miller, 2006b):

- The Central Illinois Regional Airport was concerned about the how the flashing red strobe lights would interfere with pilots arriving and departing from the airport. Horizon appeased the airport's management's concerns of interference with takeoffs and landings, radar, and even local crop-dusting operations (Riopell, 2009; Shults, 2006). Horizon built the wind farms an amenable distance away from the airport and agreed to coordinate the strobes so they all flashed at the same time, thereby limiting disorienting distractions and immediate dangers to pilots.

- "The U.S. Department of Defense and the Federal Aviation Administration issued stop work orders to several wind farms around the country," including Horizon's project in McLean, "worried the massive turbines could interfere with military radar and national security" (Miller, 2006a). The two Illinois senators stepped in to remove this development barrier (*Pantagraph*, 2006).
- The U.S. Army Corps of Engineers investigated Horizon, and some other wind farms under construction, to guarantee the projects were not harming wetlands (Miller, 2006c). During the investigation, Horizon was limited in what it could do on-site until the study was complete. Horizon was cleared to fully reengage with the McLean wind farms a few months later.

Despite the stops and starts, Horizon was on a path toward completing their Twin Groves wind farm projects by 2008.

Build Out: Planting Turbines, Producing Commodities

The phase of wind energy development in which impact is most apparent is the construction phase. Table 2.3 lays out the range of technicians and laborers employed to construct a wind farm. Massive pieces of capital infrastructure are produced and shipped across road and rail. Cranes and trucks (semis, cement mixers, pickups, and gravel) dot the landscape and crowded roadways. Restaurants and hotels are packed with an influx of temporary trade workers. Horizon actively attempted to hire as many relevant laborers as possible from the central Illinois region, but in the end, many laborers came mostly from a construction firm out of Terre Haute, Indiana, specializing in wind farm construction. So while money was spent locally, many of these laborers came from other communities, shipping a sizable portion of their income outside of central Illinois to their home communities.

The labor and equipment necessary to construct wind farms are significant, requiring coordination from many indirect actors. Public safety officials got involved to assure safe transit. Public works agencies assessed the capacity of the roads to bear the weight of the infrastructure. Many of the leases are on agricultural property accessible only via township roads, which are known for being paved with an annual application of gravel and road oil. Horizon worked with elected local township officials and agreed not only to repair the roads they damage but to upgrade those frequently used for the wind farm. A local school administrator noted that the township roads are in some of the best shape he had ever seen (Nov. 11, 2001)—one can easily see the improved roadways touched by Horizon, many of which appear to

Table 2.3. Types of direct employment from project development and construction

Backhoe operators	Interconnection labor
Clerical and bookkeeping support	Maintenance mechanics
Concrete-pouring companies	Microelectronic/computer programmers
Construction crews	Operations and maintenance personnel
Crane operators	Road builders/contractors
Developer's construction management	Site administrators
Developer's legal team	Site safety coordinator
Earthmovers	Site/civil engineers
Electricians	Tower erection crews
Environmental and permitting specialists	Truck drivers
Excavation service labor	Utility and power engineers
Field technicians	Wind energy project developers
Geophysical/structural engineers	Wind farm operators

Source: Loomis and Carter, 2011.

be of higher quality than state highways. Said a local farmer: "The turbines created nice access to the farm fields. They built nice roads!" (Dec. 1, 2011). The investment in local infrastructure certainly engenders goodwill among local officials responsible for the roadways and the farmers who use them on a regular basis.

Access to the building site requires not only the landowner's approval but also the approval for right-of-way access across adjoining neighboring properties. Cranes and other pieces of construction equipment are massive, needing to maneuver across large tracts of land. "Good neighbor agreements" secured access to adjacent land. As mentioned above, such agreements in McLean were priced at $1,500 per year, though Horizon sets the prices on a community-by-community basis, meaning Horizon is flexible on what it may pay (wind developer, Nov. 8, 2011).

We know that, as a wind energy company speculates on a given area, the process of development is long and arduous. Wind energy developers seek enabling forces early and often. In this way, wind energy companies interface with the local communities, specifically the formal local branches of government and the landowners. In total, Horizon would spend over six years preparing the project. Build-out drew in hundreds of temporary laborers. Mobilization required the organizing of crucial landowners, state, and regional regulators, and the final cost added up to more than $500 million of sunk capital. But by 2008, Horizon had completed the build-out of 240 wind turbines capable of generating 396 megawatts of electricity at peak capacity. And McLean County entered into a new community partnership for the next thirty to fifty years.

Reaping the Harvest:
The Development Implications of an Operational Wind Farm

An immense amount of financing and human capital is infused into the project during the planning and construction processes. While the wind farm itself cost half a billion dollars, local labor accounted for an estimated $50 million (Loomis & Carter, 2011, p. 23). Keep in mind that this labor force is fleeting, lasting the duration of construction. Throughout the process, the developers interact with many community groups in what appears to be a more or less predictable manner.

An operational wind farm by many accounts is an anticlimactic experience. Scholarly or journalistic accounts of wind farms and their longitudinal operational impacts are lacking and would be useful. McLean was, after all, left with a relatively docile set of 240 immovable turbines. What occurs after the construction? Is the community contractually predetermined to maintain a passive voice?

Interviews conducted during the study of Horizon revealed that many of the public officials who participated in the wind energy development process were by and large relatively dismissive of the wind farm having any profound community impact. A local economist with the McLean's economic development agency, when asked about the long-term effects of the wind farm, responded: "It's really a rural development-type thing. It doesn't do much. It's a brief shot in the arm. Just another industry coming to the area." A local farmer hosting a wind turbine said: "It's a nonevent" (Dec. 1, 2011).

There was no substantive discussion of wind energy development reducing chronic social problems or benefiting the marginalized within McLean County. Criminal activity, minority populations, the poor, youth, and elderly were never mentioned with relation to the wind farm unless prompted. Even then, interview participants noted local community-based, social welfare organizations such as the United Way, MarkFirst, Project Oz, and Sweet Home Ministries as the groups most appropriate to address such issues. Arguably, the wind farm made things worse for some populations. An economist from the economic development council noted: "There is an issue locally with poverty. But because of the countywide wealth, McLean qualifies for very little from the state (we are a $9 billion economy). Since we don't get the basics available to other counties, it makes poverty worse" (Sept. 5, 2011).

Horizon developers had no involvement with these groups (Nov. 9, 2011). The lack of participation with marginalized populations begs the question of boosterism, or why people are so supportive of wind energy development. But even beyond Horizon, it seems as though the development emphasis in McLean has more to do with growing its way out of various social ills through market development than intentional community development aimed at addressing these heady topics. The mind-set is summed up as "some development is better than no development," though some measure of local economic impact certainly helps to make a case for IOU wind energy development.

Tax revenue was frequently mentioned as a significant benefit to the area. The wind farm has benefited various taxing districts within McLean (fire protection, public library, townships, community college, parks, and county). Altogether, McLean's wind farms—not just Horizon—have created a taxable $600 million property base generating over $4 million in annual tax revenue that previously did not exist (school official, Nov. 11, 2001). The local economic development council was able to create a microenterprise loan fund from extra county tax revenue generated.

The wind farm subsumed a significant share of the area property tax burden, thereby lowering existing resident's total tax bill. Tax rates decreased for residents in these areas, and the estimated assessed evaluation of the property hosting the Horizon wind farms increased dramatically (from $60 million in 2008 to $105 million in 2010); wind turbines seem to either do nothing to property values or increase their worth despite what some opponents claim (Carter, 2011).[9] That means the wind farms have added fiscal value and have taken on a greater share of the community's tax burden, no doubt a benefit to property owners as well as local governments, though the benefits to prospective low-income home buyers, renters, and non-property owners must be low to marginal, and the urban taxing districts benefit not at all.

Representatives of educational institutions are influential boosters of wind farms (Coulter, 2009). Indeed, educational institutions in McLean may be the biggest beneficiaries of the passive wind energy development. The area's community college, Heartland, started and runs a turbine tech program with the intent of creating "green jobs." Heartland has teamed up with the local building trade groups to enhance the region's workforce capacity for wind energy development.

Illinois State University has cultivated a healthy relationship with Horizon due in part to the close ties of one of the university's prominent professors to the wind energy industry. The professor was able to translate his political capital with the DOE to start the Center for Renewables via a large seed grant supported by Horizon. The center not only works with the industry but also educates the general public on matters of wind and solar energy (association official, Nov. 3, 2010). The website for the center claims three major functional areas:

- to enhance the renewable energy major at Illinois State University;
- to serve the Illinois renewable energy community by providing information to the public, and

- to encourage applied research concerning renewable energy at Illinois State University and through collaborations with other universities (Center for Renewable Energy, 2013, n.p.).

The center does this by organizing thematic conferences, taking policy makers to wind and solar production facilities, and producing reports and research. The center also collaborates with Western Illinois University's Illinois Institute for Rural Affairs in building a wind energy curriculum for teachers across the state, setting up weather stations to assess the viability of a wind turbine, locate grants for renewable energy development, and help site potential turbines on the grounds of K-12 schools. But unless the Center for Renewables is actively discussing how to harness wind energy development for community benefit, then it is probably doing more to advance IOU interests than community interests.

Higher education institutions receive a bulk of their government funding from state sources. Local K-12 schools, primarily funded by property taxes, stand to benefit most from wind energy development. And nowhere is that truer than in the case of the rural Ridgeview Community Unit School District no. 19. An administrator for the Ridgeview district was emphatic about the material benefits of the wind farm. A former school board member claimed this district received the lion's share of the benefits from the wind farm. Indeed, no other government institution received as large an increase in revenue as Ridgeview. A study by the Center for Renewables (Loomis & Aldeman, 2011, p. 16) found Ridgeview received an $800,000 bump in total annual property revenue despite a decrease of $754,779 in state aid for the three years before 2011. "Horizon has been great to work with. Athletics are big in these rural districts. We have been able to expand our educational programming and keep athletics. For a rural school district, we're doing better than a lot of the others" (school official, Nov. 11, 2001).

These passive development outcomes are a result of existing local structures that seek to benefit from the predictable market activity. The mere existence of the wind farms has stimulated local public investment in the status quo and strengthened the existing local institutions. Higher education facilities are taking advantage of having accessible facilities nearby through the creation of new programs. Government taxing bodies are best positioned to take advantage of Horizon's presence in the area. Officials with these local taxing authorities are pleased with the influx of additional revenue, particularly during an economic downturn that has decreased the revenue of their counterparts that are not playing host to such infrastructure. Fire districts aren't downsizing, and the township roads are in the best shape they have ever been. Plus, the Ridgeview school district is expanding rural education when national trends are encouraging austerity and consolidation. However, the direct impacts of Horizon's activities on local community governance are less clear, as examined in the following section.

Community Participation or Roadblocks?

Horizon, like many other energy firms, is multinational. These companies develop partnerships with other major corporations for financially beneficial relationships. These relations come about due in part to federal and state subsidies that incentivize such partnerships. But interactions go beyond this to other groups within the local community.

A pattern culled from the interviews with community members is the seeming disappointment in economic outcomes, particularly concerning employment. Many of the interview participants noted that job creation was much lower than anticipated, and a few cautioned that other communities should be made aware of that fact. During the fieldwork for this study in McLean in 2011, Horizon went from two developers to only one on staff who divides his time among new projects in Ohio, Indiana, and Illinois. Horizon indirectly employs a local workforce, contracting out through General Electric to provide maintenance services. According to the developer, Horizon typically has thirteen General Electric workers on contract.

Horizon's active involvement in the local community life is nonexistent outside of furthering wind energy development interests. Horizon is typical of wind energy firms: it seeks out communities in order to extract value for their investors. Wind energy firms are positioned best to "sell" their project on a fiscal basis. The ability to talk dollars and cents and rapidly infuse capital means their value proposition synchronizes well with communities playing host to a growth coalition. And that makes sense when one considers Horizon and the other actors in McLean.

McLean exhibits many of the characteristics of a growth-oriented community. Governance regimes appear to be well established. The chamber of commerce, economic development council, planning commission, farm bureau, and city and county governments were often cited as the enabling forces within the community. These institutions are all inclined to support economic development endeavors with little regard for the broader social dimension as long as lucrative property development and fiscal streams appeared a likely outcome. Development is geared toward standard material aspects of life. Jobs were persistently noted as important policy outcomes, and none of the discussions brought up a desire to empower the marginalized or even mentioned their existence.

The wind farms did more to enable this status quo than to open existing social systems or build new institutional access points. Material benefits mostly went to landowners with significant plots of land; the total annual lease payments amounting to over $1 million. The landowners who are party to such lease agreements are more often than not row-crop farmers with a well-protected territory and significant influence in local institutions.

Surprisingly, the major corporate interests in McLean had little to do with the growth coalition. A number of the interviewed residents noted that the corporations and associations headquartered in the county were important regarding contributing to local civil society functions. Companies were known for allowing and encouraging their employees to do volunteer work, as well as nonprofit or government board work, on paid company time. When asked about the type of work, many respondents were vague about the emphasis, but a lot of their efforts appeared geared toward professional, resume-building endeavors in the market (volunteering with the local hospitals came up often) and, to a lesser extent, the civil society realm.

It was paradoxical that in many of the interviews, when the question was asked, "Who would I contact to get things done or to block a project?," the responses never mentioned the large firms headquartered in the community. Should they desire, these firms could wield significant local influence, but it makes sense that they do not, probably because their areas of emphasis are most often outside of the community, focused on their core competencies at a broader socioeconomic scale. From this perspective, McLean County becomes just another place to do business, not necessarily a community necessitating economic contestation. While some of the professional staff from these businesses may be involved in the growth coalition, their institutions are not positioned to govern locally over community life. Indeed, should these corporations face local conflict, they could either threaten to leave the area along with their economic clout, or "level jump" to change the regulatory environment in their favor, a practice more in line with corporate, institutional logics.

Considering Horizon's Institutional Logic

The wind farm certainly enhanced the fiscal position of local landowners and critical local taxing bodies. Horizon's benefits to a select few well-established property owners (330 property owners signed leases, good neighbor agreements, and easements; Loomis & Carter, 2011, p. 16) and the local social structure controlled by a growth coalition do not mean the development outcomes are negligible or even undesirable. Local taxing bodies, particularly K-12 school districts, were primary beneficiaries, allowing public services to avoid austerity measures during the Great Recession (in most instances, the services were enhanced).

Yes, the development of the wind farm is of a more passive variety, stemming from the wind farm being embedded within many institutional arrangements

(FERC, MISO, finance, etc.), thereby hyperfocusing on the operational features while spreading thin the local community focus. Recall that, for the wind farm to be marketable, it must connect with the national electric grid. The utilization of PPAs as a standard for bringing wind energy to market indeed is a benefit to consumers in that it decreases overall price volatility (association official, Nov. 3, 2010). And one would be remiss not mentioning the obvious: wind energy production does not pollute. For every wind farm, the need to utilize carbon-emitting power plants diminishes. Those situated within the lower economic strata—the working poor and individuals impacted by environmental classism and racism—would no doubt find these incremental material developments to be a positive outcome.

Wind energy offers broad socioecological benefits beyond greenhouse gas reduction. Horizon's wind farms in McLean will make it harder for sprawling strip malls to develop over dark, fertile soil (no one wants to build parking lots underneath wind turbines, especially since in the winter turbine blades are known to fling large icicles). Wind farms may then serve the purpose of fighting urban encroachment while sustaining rural landscapes through the formation of green belts. In this way, wind farms could be used as a market-based approach to regulate land use, save green spaces or farmland, and contain sprawl. But for this to occur, development must be intentional. Local policy makers and leaders cannot presume that the mere construction of a wind farm will result in some positive spin-offs; they have to make it happen by design.

Surprisingly, environmental impact was never mentioned as one of the development outcomes in interviews, despite the environmental angle being at the forefront of the wind energy developers' marketing campaigns. A state regulator summed up his take on wind energy simply while rubbing his fingers together: "Money. This is big business. It ain't as cuddly as they want you to think it is" (Oct. 25, 2010).

The wind farms also brought out a new arena for contestation in the community. The opposition, mentioned only by three interview participants, was dismissed as misguided and somewhat disruptive: "It's too bad that wind farms can divide a community" (business executive, Dec. 1, 2011). Another interview participants dismissed opposition figures as inexperienced and out of the local norm: "There really isn't what you might call an activist type crowd here" (association official, Nov. 11, 2011).

The standard defense among these interview participants was essentially, why not develop a wind farm, what's it matter to them? But the perception of the opposition seems to have been shaped from observations of the one-way relationship between key community leaders who supported the project and the Horizon developers. In an interview, a Horizon developer openly noted that the development team utilized tactics to identify their support base and subvert opposition: "Long-time residents typically have no problem with turbines. Long-standing locals are the best . . . but the new folks are different to work with. It's the new rural folks, the people who treat the area like bedroom communities that we have a problem with" (Nov. 8, 2011).

According to the developers, residents who spoke out against the wind farms typically were not farmers, owned small one- to two-acre parcels, and were not generating revenue from their land. Opposition figures wanted to maintain the rural character of their community, to be untouched by the overwhelming visuals of hundreds of spinning turbines (local official, Sept. 5, 2011; association official, Nov. 11, 2011).

The Horizon wind developers have done their best to avoid the smaller villages where opposition actors were mobilized and instead directed their attention toward those locales with a strong support base. The opposition saw these approaches as subversive. But the opposition narrative was mitigated by the control of crucial governance institutions by wind energy supporters and the opposing community's lack of long-standing civic action and mobilization around big-picture issues. A persistent, successful growth coalition can build broad-based acquiescence. Over time, people forget how to do opposition, to engage in collective action. Over time, this ends up as economic and environmental justice denied.

As wind energy infrastructure occupies more and more of the rural landscape, groups such as Information Is Power will increasingly challenge wind energy development through many formal procedural mechanisms (Brady-Lunny, 2007). The rise of exogenous opposition groups influencing local collective action may be beneficial to civic life in a place like McLean in that these external groups may reorient people toward working collectively for mutual ends outside of the dominant social system. An outstanding question is whether or not those groups can convert from an oppositional force to one furthering civics and community development.

Importantly, the spatial layout of wind energy infrastructure has the potential to democratize energy governance. The sprawling nature of the infrastructure means that, unlike a centralized coal-fired power plant, a wind farm will interact with far more people in its host community, at least in the development stage. Additionally, consider that the deployment of clean, renewable energy is of increasing necessity. Recent research has highlighted the rapidity of global climate change; we have a limited time frame with which to reduce or eliminate greenhouse gas emissions (McKibben, 2007). But as the situation in McLean seems to demonstrate, systemic problems of inequity may be one of the most significant impediments to deploying wind energy. If people don't have some sense of ownership over the energy development and climate change processes, we can expect a lack of civic engagement or outright hostility toward the perceived enrichment of a few at the expense of the many. This is where the democratization element comes into play.

The Horizon wind farm is not as much a mixed bag of new economic development approaches as it is a continuation of old growth models and mild strengthening of the local status quo. Horizon is not entirely transparent in what it can do for a community unless it is asked (and community leaders have to know the right questions to pose to uncover this information). Illinois installed more new generation capacity than forty-eight other states in 2011 (Loomis & Carter, 2011), and Horizon looks to be a big part of future growth in Illinois (Miller, 2007b; Sapochetti, 2010). Considering the number of subsidies that are injected into investor-owned wind energy for private gain, Horizon and other investor-owned generators bear a public responsibility to speak openly with their host communities about the range of development services and outcomes they could offer. But unless these companies are prompted, chances are they will not readily divulge such information.

It is questionable whether an investor-owned wind farm, under current multi-level government policy arrangements, can contribute much to a community outside of material enhancement. Granted, the material element may be to increase funding to local community development groups, but these interests will always be at tension. Community development will ideally break dependency, whereas a profit-seeking firm will seek to maximize political advantage and value extraction, meaning that deepening dependence on the investor-owned firm may be a predetermined design feature that is nonetheless detrimental to community governance.

There is, then, a built-in, self-replicating dependency mechanism that bolsters the importance of large firms like Horizon. Community reliance on these types of firms could be disastrous if national or global markets collapse. Part of the reason these firms can contribute so much human capital locally is because of the wealth they extract from other areas and concentrate back to their central headquarters. In an era of state austerity measures, it is not far-fetched to consider that a perfect storm of state failure might occur in which promised subsidies are not paid to the wind energy company, thereby destroying the firm's bottom line. When IOUs stop producing revenue, what options does the community have to weather such events?

It appears that, if left to the desires of the wind energy firm, it would seek the path of least resistance toward the end goal of a fully operational wind farm. This lowest-common-denominator tack runs the risk of excluding marginalized segments of the local community, driving further divisions into the local social fabric. What is more, without active input from the local community, specific concerns may not be addressed and accounted for (e.g., the possible utilization of the wind farm to curtail urban sprawl onto local farmland). The community simply must be involved in conceptualizing long-term ramifications and participating in the planned build out to maximize the public good created by a new wind farm.

Uncertainties remain about how to influence development outcomes from wind energy. Dependency pitfalls arise when community stakeholders lean on federal and state policy makers to create criteria for the wind energy companies to receive various subsidies. Perhaps opposition or social justice groups at the community level could mobilize to better guarantee that host communities are not exploited and receive a just share of newly created value.

Attention now turns to the cooperative business model of wind energy ownership.

CHAPTER THREE

Case Study—The Cooperative-Owned Wind Farm

Investor-ownership dominates wind energy. As of 2013, there were only a handful of community-owned utility-scale wind farms. These wind energy organizations are relatively isolated, with the lack of linkages making it difficult for actors to pool resources and grow community ownership. That means policy makers at the local, state, and national levels have relatively few examples with which to assess the ideal institutional arrangements possible to optimize community economic development outcomes from public investment in wind energy.

As shown in Chapter Two, investor-owned wind energy continues to operate on the logic of wealth extraction. The investor-owned firm requires significant monitoring by public interests to protect consumer and community members from its entrenched extractive orientation. Perhaps it is time to stop privileging with public monies the investor-owned firm, which seeks to restrain its community responsibility.

What could, or should, be expected of a wind farm, particularly in terms of how the organizational structure affects its interaction with its host community? Since wind farms are heavily government subsidized, the argument exists that wind energy firms benefiting from subsidies should contribute back to collective community well-being.

These questions are what make PrairieWinds ND 1 Inc. in Ward County, North Dakota, of particular interest. The arrangement serves as an outlier because Prairie-Winds, a subsidiary of Basin Electric Power Cooperative that was opened in 2009, is the nation's first utility-scale cooperative wind farm.[1]

The question of what happens to a community playing host to the nation's only cooperatively owned wind farm is interesting in part due to the institutional design of cooperatives. The defining feature of the cooperative ownership model is that a cooperative is "an autonomous association of persons united voluntarily to meet their common economic, social, and cultural needs and aspirations through a jointly-owned and democratically-controlled enterprise" (International Co-operative Alliance, 2015, n.p.). Chapter One addressed the rationale (institutional logics) for the efficacy of the structure of the cooperative as a community economic development institution: the features of a cooperative provide a venue for assembly, fostering social capital through civic interaction. The cooperative business model thus instills self-governing, democratic values into its membership and partner organizations. The orientation of the cooperative business model parallels the ODPs for enduring sustained collective action through robust institutional arrangements, mitigating disempowerment, alienation, and dependency-building mechanisms of monocentric systems and enhancing the potential for polycentric self-sustaining, self-governing institutions.

All utility-scale wind farm developments are major, capital-intensive projects; they draw in hundreds of workers, cost millions of dollars, and change the face of the rural landscape. Any utility-scale wind energy development will cause appreciable transformation within a host community. But interactions among the ownership structure of Basin, government and market incentives, and the socioecological features of Ward County, North Dakota, influence the development outcomes of the wind energy development project.

This chapter contextualizes the case: Ward County, North Dakota, and the PrairieWinds wind farm. While Chapter Two focused on understanding how the IOU-owned Horizon wind farm operated within the electric energy system and its implications for the community of McLean County, Illinois, the cooperative is the institution of interest here, which is occasionally be contrasted with IOUs. This is because electric cooperatives operate within an IOU leveraged system and must be understood from that vantage point. The chapter then addresses the development processes and governance initiatives that made the wind farm a reality, provides an analysis of development implications, and concludes with a discussion of the research findings surrounding PrairieWinds.

Ward County, North Dakota: Robust Organizations Within a Boomtown

At the time of the fieldwork in 2011, two things stood out about Ward County: it had been an oil boomtown county since the mid-2000s and during flood-ravaged 2011. As a lifelong resident described: "The community before the oil boom was more stable, its focus was more rural, more on agriculture. Ag-related processing (a mill for pasta, another place for beans, lentils) really drove the area" (Dec. 8, 2011). Another lifelong resident noted the rural proximity of Ward County: "The county is very rural and urban at the same time and touches the Mandan, Hidatsa, and Arikara Nation on its southwest border. It had always been a good place to live, start a family, and have a career" (Aug. 8, 2011). Steady economic growth meant that people could count on an agriculture-driven economy that provided for individuals and their families.

Enduring civic cultures coupled with the brutal seasonal swings in weather instill a sense of collective responsibility among North Dakotans. One prominent statewide economic development professional stated that, "if you're pulled over the side of the road, I guarantee that the first person who comes upon you is going to pull over and help you. No one wants to be left to fend for themselves when they're stuck in a ditch during the North Dakota winter" (Aug. 8, 2011).[2] The same economic developer spoke to this in his own experience with flooding in nearby Bismarck:

> The river was coming up over the bank right toward our house. We busted our tails to put up a makeshift levy. But—and this speaks to the character of the people in these parts—some of our neighbors showed up with a semi full of city sand and a backhoe, without being asked! He saved us hours, maybe days of work. And he refused to be paid! He just wanted to help out his neighbors. But that's not all. The Bobcat Company lent out tractors and loaders to use to fight the floods . . . for free. What other state do you know that has that sort of social capital? (Aug. 8, 2011)

When looking for the loci of activity in Ward County (population 61,675), one necessarily turns toward the county seat, Minot, the largest city in Ward and fourth largest in North Dakota, at 40,888 total (Minot Area Chamber of Commerce, 2013) (the mayor believes Minot's population has exceeded 50,000 since the 2010 Census; Zimbelman, 2013). Minot is where people meet many of their day-to-day needs and serves as a central hub of vital collective pursuits.

The last few years have situated much of North Dakota—and particularly Ward County—firmly within a maelstrom of global forces (Schramm, 2012). Agricultural markets and federal subsidies are enriching local farmers who are adapting to new crops (corn is replacing lentils and sunflowers in some areas) and farming practices. The energy boom has brought in a great deal of in-migration and economic development. Long-term residents expressed concern and anxiety that things are changing in ways that will forever reshape the uniqueness North Dakotan culture (local residents, Nov. 1, 2010, and Dec. 8, 2011).

The Minot Air Force Base (MAFB) has done its part for economic development. The U.S. Department of Defense (DoD) allocates resources spent locally by service women and men, as well as the support personnel necessary to run the facilities that traditionally served as the home of a B-52 bomber. In a sign of the times, MAFB is one of the command centers for the DoD's unmanned drone warfare program. As drone warfare expands, so too will the personnel at military bases like MAFB. Expansion of DoD resources at MAFB would typically be a remarkable story on its own. However, the shale oil boom over the Bakken and Three Forks deposits creates a lot of noise, drowning out almost everything else.

The Bakken and Three Forks oil deposits are thought to join up with lower Canada, and reach through northwestern North Dakota into parts of Montana. Development of these resources seems to be taken for granted, along with a change in local culture, as expressed by a local: "North Dakota will become the next Texas. We have so many energy reserves that it will reshape this state." He then went on to caution: "And I don't know if it's for the better" (cooperative official, Dec. 8, 2011).

The boom touches virtually all of the communities within proximity to the oil deposits. According to the Minot Area Development Corporation (MADC), exploitation of the shale has turned North Dakota into a net energy producer, sending 75 percent of energy reserves out of state (Minot Area Development Corporation, 2013c). The rush for shale-enriched land by companies such as Chesapeake Energy, Haliburton, and Marathon has attracted a great deal of investment, heavy equipment, and migrant laborers. While the oil deposits mostly surround Ward County, Minot's position as a regional hub makes it an attractive area for energy-related business activity, as well as housing and recreation for the labor force.

A persistent narrative running through the interviews was the extent to which key leaders in the region were ill-prepared for the boom. The sheer volume of migrant labor and heavy equipment has taken a toll on the region's infrastructure and service industries. Sewer and water lines are increasingly stressed. Montrail Williams, a rural electric cooperative servicing surrounding counties, was at one point stringing up three miles of electric line daily to meet the needs of the energy companies (Schramm, 2010). The country roads, originally built to handle the occasional piece of farm machinery, are now dealing with an endless barrage of semitrucks. "The driving is difficult. Some days just trying to get off the bypass seems impossible" (local resident, Dec. 8, 2011).

A local community leader noted: "The country roads went from asphalt to mud. The truckers are driving their payloads long distances at fifteen to twenty-five miles an hour because the asphalt isn't there anymore. But that's just the thing. It doesn't matter if the roads are paved or not. Nothing's going to stop them from transporting the oil. It's literally 24–7, rain, snow, or shine" (school Official, Aug. 16, 2011).

The area is underequipped to deal with the rapid increase in trucking. Truck stops are at capacity, and staffing is not adequate; there are stories of truckers being handed buckets to use as makeshift toilets while overnighting. Even the harsh winters seem to cede to the oil boom: "During the snows, people stayed in more. Outdoor-type things were at a standstill. With the boom, people are more involved outdoors during the wintertime" (local resident, Aug. 10, 2011).

Auto repair shops are exceeding capacity to meet the needs of locals. Mechanics are servicing not only long-standing residents but also the rigs and pickups used for working on the shale. The result is that regional auto shops are servicing Ward

County as well. "If I need a car repair, it's just easier to pay to have it hauled eighty miles away to Bismarck. I'm not gonna wait weeks to be able to drive again. You need a car in these parts" (local resident, Dec. 8, 2011).

Nationally, antifracking tales are being told through widely distributed documentaries like *Gasland* (2010), as well as highly organized environmental campaigns. However, the companies operating in North Dakota have gone relatively undisturbed in their business operations. It was only recently that state governments restricted the oil companies in what fluids they were able to use for fracking. However, that was not born out of opposition as much as a proactive initiative by the oil companies to shield themselves from future litigation (*Minot Daily News*, 2012b).

Perhaps the economic impacts of the oil boom have mitigated fallout from the pitfalls of the boom. The influx of external investment from the oil and gas industries has brought with it an employment boom that seems to have shielded North Dakota from the Great Recession (Table 3.1). Minot economic indicators drive home the extreme economic impact the oil boom has had on the community, represented in hefty tax receipts and extraordinarily low unemployment.

The high rates of employment do have downsides for economies reliant on abundant wage labor. The oil boom has resulted in a massive labor shortage throughout the region despite the rapid in-migration of laborers. Service and support job openings are plentiful. These new laborers are in need of essential services, stimulating a boom in retail industries such as grocery and restaurants. Businesses are finding that the regional labor force is in demand, meaning workers can job-hop, increasing workforce turnover (business owner, Aug. 16, 2011). In the midst of this uncertainty, business owners of traditionally low-wage jobs are offering pay rates upward of $15/hour, signing bonuses (which increases job-hopping behavior), and higher benefits packages (including health care and vacation time) no matter the prospective employee's level of education or experience. "It used to be that the jobs I would offer in manufacturing were the ones people wanted. Now I have to compete with the McDonalds on wages and benefits. I just don't see how we're going to adjust to this mess. It's a real strain on my resources" (business owner, Aug. 16, 2011).

Many fast food restaurants are so understaffed that they have closed down their indoor seating and service only the drive-through window. A chain housewares retailer "is a madhouse" (cooperative official, Dec. 8, 2011). Even Walmart has to adapt, pulling pallets of product onto the sales floor instead of stocking items on the shelf due to massive product turnover and limited staff.

Mass in-migration has resulted in a regional housing shortage. The oil and gas companies have set up temporary labor or "man" camps for their employees. Laborers willing to pay for housing outside of the camps have occupied virtually all of the

Table 3.1. Minot ND economic indicators, 2008–2012

Indicator	Year				
	2012	2011	2010	2009	2008
Workforce[a]					
Minot employment (Ward County)	29662	29550	29280	28695	28544
Minot unemployment (Ward County)	2.80%	3.60%	3.60%	3.80%	2.90%
North Dakota unemployment	2.80%	3.50%	3.80%	4.10%	3.10%
U.S. unemployment	7.60%	8.90%	9.60%	9.30%	5.80%
Minot city sales tax collections[b]					
Yearly total	$26,285,298	$20,828,684	$17,200,990	$15,112,850	$14,242,642
Taxable sales and purchases[c]					
Minot	$434,443,169	$1,443,645,118	$1,072,382,163	$906,784,178	$841,067,276
Ward County	$451,426,155	$1,505,846,697	$1,124,212,220	$952,338,763	$885,195,620
Building permits and valuation[d]					
Total units	1,419	1,132	652	400	366
Single-family homes	363	292	138	151	151
Multifamily units (apartments, etc.)	1,056	840	514	249	215
Residential valuation	$147,205,336	$88,325,200	$52,021,800.00	$34,583,900	$39,362,518
Commercial valuation	$157,742,100	$116,235,000	$48,181,000	$31,378,300	$40,888,000
Total valuation	$304,947,436	$204,560,200	$100,202,800	$65,962,200	$80,250,518

Table 3.1. (*continued*)

Indicator	Year				
	2012	2011	2010	2009	2008
Cost of living index[e]					
100% Composite Index	107.2%	103.6%	99.9%	97.1%	95.5%
Grocery items	101.4%	96.8%	99.3%	99.9%	100.9%
Housing	126.5%	107.3%	95.6%	90.4%	86.8%
Utilities	73.8%	76.2%	73.7%	71.9%	89.2%
Transportation	111.5%	98.3%	98.2%	97.6%	94.6%
Health care	108.4%	100.2%	90.8%	92.8%	91.8%

Data are annual figures, except as stated below.

[a] Data for 2012 are for December only. Unemployment figures are not seasonally adjusted; there were 2,057 job openings in Ward County in January 2013. Source: Minot Area Development Corporation 2013a.

[b] Source: State Treasurer's Office.

[c] Data for 2012 are for the third quarter only. Source: State Tax Commissioner's Office.

[d] Source: City of Minot.

[e] Data for 2012 are for the third quarter only. Source: American Chamber of Commerce Researchers Association.

existing hotel rooms in the region, necessitating a growth in hotel construction (ten hotels were being built as of the time of this research; cooperative official, Aug. 15, 2011). However, many of these laborers are men, mostly young and single, working overtime with excellent pay packages; they can afford to pay inflated prices for the enhanced amenities incurred from city life.

Speculation on development property is occurring as the willingness of prospective tenants to pay inflated rents increases (school official, Aug. 16, 2011; local resident, Dec. 8, 2011). Increasingly, out-of-town investors are seeking to profit from this. Interstate landlords are purchasing local apartment complexes and converting hotels to long-term, temporary-stay facilities. The result? Rent and property values are skyrocketing. A journalistic account finds rents on par with New York City rates (Johnson, 2012). The willingness to pay a premium for standard housing in Minot is driving up the value of the entire housing stock, inflicting pain on the pocketbook of long-term residents as the cost of living has increased. "The people who are local are having big problems with landlords who are jacking up rental prices. The homelessness is a different problem. But people who have good jobs are unable to get a place. One-bedroom apartments are going for around $1200 a month" (cooperative official, Dec. 8, 2011).

A resident gave an example of an apartment complex that was purchased by an out-of-town corporation. The corporation sent notices to current occupants that the rents would triple. He went on to wonder, "What happens when this boom is over, and the locals have been forced to move out?" Retirees and other individuals on fixed incomes living in rental properties are being squeezed out of the housing market, and the increased wages are sapped to cover additional cost-of-living expenses.[3]

Rapid in-migration is beginning to build mistrust in individuals as well as local institutions, another indication that the community is going through rapid expansion: "The influx of people has pushed an increase in police calls and things like that. People are concerned about it, and you hear rumors about things that happen in the parking lot of Wal-Mart, but the police department downplays it." While crime data are difficult to find, one website notes a significant uptick in violent assaults in Minot. More recently, the state's attorney general released a report that concluded increases in crime rates were negligible (Preskey, 2012). However, enhanced fear of outsiders and the perceived increased potential for random criminal victimization will make collective action more difficult due to diminishing trust. This is troubling for a community enduring rapid transformation; stable, long-term planning will be critical to ensure a smooth, stable transition during and after the boom.

The spring of 2011 saw record rainfalls hindering local crop planting. The rainfall poured over already saturated ground, causing the Souris River to rise over its banks. Initially, a few essential interstate highways were closed down due to flooding effects on structural integrity and transit safety. But the rains kept coming. Large-scale flooding hit the region. The record flooding of the Souris River "left about 11,000 people—more than a quarter of Minot's population—effectively homeless" through the destruction of over four thousand housing units (Bailey, 2011, n.p.). The flooding had compounded the chaos in Ward County: "There isn't any place to stay in Minot with 10,000 people displaced by flooding and with the oil workers occupying every hotel room in town" (local resident, Nov. 1, 2010). The oil reserves driving the boom had always been located outside of the flood zone. The flooding left the energy boom to continue unabated, which only compounded the problem of accessible housing for newly homeless residents. A housing report before the flooding estimated that Minot would need some five thousand new homes by 2023 (Ondracek, 2011). Now Minot would need to replace four thousand homes from its existing housing stock, a virtual overnight doubling in demand. The flood had also hit the region surrounding Minot, compounding resource shortages. "In Burlington, a town of about 1,000 people a few miles upstream on the confluence of the Souris and Des Lacs Rivers, city officials abandoned sandbagging as hopeless and sent people to Minot to help out. About a third of the town's 320 houses are expected to be lost" (Kolpack, 2011, n.p.). The flooding of the region's Amtrak rail line further constrained newly homeless, carless residents from leaving the area (Minot Recovery Information, 2013).

The disaster recovery process itself further complicated the housing shortage. Many of the properties damaged within the Souris River flood zone are in a holding pattern. Minot city officials must decide if the area is safe for continued residential zoning, or if the properties should be converted into a green zone or park to mitigate the impacts of future flooding on residential housing stocks. "Minot is trying to come up with a plan for flood protection which involves property buyouts. People are not rebuilding their homes until they find out what happens" (cooperative official, Aug. 15, 2011). Since the close of the fieldwork phase of this research, Minot has implemented a voluntary property buyout plan, utilizing over $60 million of state government monies.

The Federal Emergency Management Agency (FEMA) is actively involved, funneling critical federal resources and locally setting up temporary housing units. However, these housing units went severely underused. A significant number of residents felt uncomfortable moving into government-provided housing and instead moved in with friends and relatives (cooperative official, Aug. 15, 2011; local resident, Dec. 8, 2011). It was not uncommon to see tents pitched in the yards, sheds, and garages of Ward County residents and occupied by flood victims.

Disaster recovery efforts have further escalated the demand for laborers in the building trades. Tradespeople are increasingly difficult to come by. Their skills are needed to not only rebuild Minot but also to service the oil boom. (Minot's city government has issued building permits for projects valued at $100 million in 2010, $200 million in 2011, and $300 million in 2012—rapid growth in built capital; Zimbelman, 2013). The stock of credentialed tradespeople is being stretched to its absolute capacity in North Dakota, exacerbated by the state's licensing requirements. Even in times of crisis, out-of-state tradespeople must still go through North Dakota's credentialing process to legally practice their craft within the state. The pressure on the state has sped up the licensing process for out-of-state contractors. Hundreds of contractors are coming into the area to help rebuild (the city had licensed over 650 new contractors between January 2011 and January 2012 alone): "To help put that number in perspective, the normal amount of electrical contractors in Minot is believed to be about 30. There are 170 now. The number of licensed plumbing firms has increased from 11 to 55, excavators from 10 to 105" (Fundingsland, 2012, n.p.). Even if new contractors make it to the region, they are finding living accommodations to be virtually nonexistent; many are camping or sleeping in their work trucks (local resident, Dec. 8, 2011).

The energy boom has not approached a predictable, stable equilibrium. One in three businesses is hiring, with an estimated fifteen hundred laborers needed for 2013 alone (Strasburg, 2012). Industry executives, laborers, and support teams will continue settling in the region. There is explicit concern that Ward County will look significantly different within the span of a few years. According to a prominent local Red Cross volunteer, about 80 percent of flooded residents were without flood insurance (Aug. 10, 2011). One interview participant claimed this was a result of local realtors guaranteeing prospective homeowners in the floodplain that the river's three dams would forever guard against potential flooding (school official, Aug. 16, 2011). What will happen to those who lost everything? Can the actors in the region adapt and return to a level of normalcy?

The city's resources are stretched thin. The oil money has generated tax revenue for state coffers, with incidental revenue captured by the local governments (sales and property taxes). North Dakota is one of the few states to have the good fortune of a budget surplus during the recession in the late 2000s. But the national wave of newly elected austerity-oriented politicians elected to state governments has complicated the distribution of resources to the disaster area (the state is in tax-cutting, not tax-spending, mode; local resident, Aug. 8, 2011).

The character of the community will no doubt change rapidly in the near term. The boomtown atmosphere, the escalating economic tensions on the long-term residents, and rapid in-migration are forces chipping away at the community's foundation, supplanting it with a new structure. According to many interview participants, these changes are prompting some long-time residents to leave as they seek more stable communities. "The entire makeup of the area I grew up looks different. It's changed everything. The community isn't the community I remember. We don't even want to go into Minot anymore. It's just a different place. The character of the area has changed a lot" (local resident, Dec. 8, 2011).

Local policy makers are now being forced to deal ad hoc with all of the issues of a boomtown and disaster zone. Understandably, interview participants expressed a desire for stability.[4] People are vulnerable and desperate to see an end to the crisis atmosphere. The desire for stability seems to have cemented the economic planners and policy makers as a de facto force in community vision and economic development. They have the resources, time, and community support as knowledge leaders entrusted with steering Ward County through good times and bad. In this environment, co-optation and capture by a select few is undoubtedly a real possibility.

Before the floods and before the oil and gas boom, the nation's largest generation and transmission electric cooperative (G&T), Basin Electric Power Cooperative, made a significant $240 million investment in wind energy. Typically, such a massive investment in localized infrastructure would be a noteworthy event in a relatively isolated county the size of Ward. Proponents of cooperative development point to claims of enhanced community outcomes. Prominent residents seemed somewhat dismissive of the substantive development outcomes of cooperative business development. What are the implications of cooperative ownership of the wind farm for the local community? How did this rare, cooperative wind farm come to be? Are the perceptions of cooperative development consistent with findings from on-the-ground fieldwork?

Basin Electric Power Cooperative: A Legacy of "Giant Power"

The standard IOU is profit oriented, whose goal is maximizing and returning value to the shareholder. Whoever has the most voting shares in an IOU more than likely can command the greatest voice (one share, one vote—the rule of capital). Cooperatives, unlike IOUs, are almost always operated by those who use them: the consumer member-owners. Shares of stock do not dictate control over the governing process, as every single member is granted the same statutory rights of governance (one member, one vote—the primacy of the individual).

From their very inception electric cooperatives designed member-owner governance into the model, providing venues for member-owner input and participation. The over nine hundred electric cooperatives in the United States are mostly electrical distributors, meaning they service their member-owners within their community; in this manner, the cooperative is spatially accessible (comparatively, many IOUs encompass patchwork tracts of land over regional service territories). The cooperative member-ownership could then harness surplus resources (profit, labor, or otherwise) for critical public entrepreneurial endeavors.

Despite the seemingly populist elements of electric cooperative governance, electric cooperatives did not come about solely as a result of a social movement dedicated to community ownership of electric systems. In fact, the existence of electric cooperatives is more a result community-based collective action to address the failure of IOU market actors to meet the electrification needs of rural locales than of political opposition to the status quo of IOUs. This community-need orientation has profound cultural implications within the electric cooperative community (to be addressed later).

While many electric cooperatives are relatively small compared to typical IOUs, an anomaly exists among electric cooperatives playing a major role in the lives of many North Dakotans. Nestled within the state capital of Bismarck, eighty miles south of Ward County, is the headquarters of the nation's largest electric cooperative, Basin Electric Power Cooperative. Basin is a vertically integrated power generator and transmitter (as one executive likes to say, "Basin goes from mine mouth to meter"; Aug. 9, 2011). Basin covers nine states, stretching from the Canadian to the Mexican border, and "owns 2,165 miles and maintains 2,250 miles of high-voltage transmission" (Basin Electric Power Cooperative, 2016c) through joint pacts and ownership (Basin Electric Power Cooperative, 2016d). "Basin Electric operates four baseload coal-fired power plants, and three gas combustion sites (9 turbines), two natural gas turbine units, and a two-unit, oil-fired plant as peaking units. Basin Electric owns but does not operate, 40-MW of a natural gas/oil-fired combustion turbine, also used as a peaking unit. In Basin Electric's renewable energy portfolio, we currently own and operate wind turbines near Chamberlain, SD, White, SD, and Minot, ND" (Basin Electric Power Cooperative, 2016c, n.p.). While the scale of Basin is staggering, it is the ownership and governance of the institution that makes it noteworthy. Basin is what is known in the electric cooperative sector as a super-G&T. It is a third-tier cooperative owned by other smaller, second-tier cooperatives (also G&Ts) and first-tier distribution cooperatives. These first-tier cooperatives initially joined together to form regionally based G&Ts (second-tier cooperatives). These G&Ts eventually joined together to form Basin to achieve greater economies of scale.[5] Under Basin, these networked cooperatives pool together to aggregate their resources to own and operate capital-intensive transmission infrastructure, build electric generation capacity, and increase their wholesale market power through bulk purchasing and coordinated bargaining. Basin generates over $1 billion annually in revenue, operating on an at-cost, not-for-profit basis. Profits either are retained for research and development or as self-insurance or are distributed back to member-owners as capital credits (cooperatives in other industrial sectors refer to this line on the balance sheet as dividends, patronage, or capital credits). Basin then provides numerous services for the entirety of its cooperative ownership outside of simply generating and transmitting electricity (monitoring, coordinating energy supply, marketing, and, as described below, community and economic development functions; cooperative executive, Oct. 26, 2010).

Distribution cooperatives, not the G&Ts, are tasked to interact directly with the member-owners (the consumers) through the procurement and provision of electric energy to the household. Those households, businesses, and entities that receive electricity from the distribution cooperative are considered member-owners. These member-owners then have a right to run for positions on the board of their particular cooperative and cast votes when called for. Electric cooperatives are then representatively governed by an elected few from the ranks of the member-ownership. These distribution cooperatives may purchase their energy from investor-owned generators or, ideally, own a part of a G&T like Basin that procures energy on behalf of those distribution cooperatives.

Basin is much different from McLean County's Horizon Wind Energy, LLC. Investor-owned wind generators like Horizon are heavily regulated and specialize in a singular aspect of the electric energy market: wholesale power. The power they generate is sold via market mechanisms with little to no interaction with consumers and their community (market logic dictates extreme specialization). Basin, representing the vision of a more complete cooperative electric energy system, is relatively unregulated by government actors and has the intent to operate a complete integrated electric grid. The historical trajectories that brought distribution cooperatives to existence have shaped how electric cooperatives join together to provide scale and agility in the broader energy marketplace in which electric cooperatives seem inextricably embedded. The market-based core of electric cooperatives also influences and clashes with the organizational value proposition, and in part determines how the cooperative interacts within the local community.

Market Failure, Community Gain

Electric cooperatives are typically referred to as rural electric member cooperatives (REMCs). While the rural characteristics of the sector are changing, rural remains a deeply held identity of the REMCs due in part to the origins of the sector.

During the Great Depression, only 10 percent of rural America was wired for electricity. The public policy that breathed life into the electric cooperative sector was intended to modernize and industrialize the American rural countryside. Under President Franklin Delano Roosevelt (FDR), full rural electrification was viewed as a significant complement to New Deal reforms intended to jolt the economy back to life. It was believed that if rural America were modernized through electrification, innovation and industrialization would progress, which would enhance rural entrepreneurship and spill over into the broader national economy (Davis, 1986, pp. 491–492). The challenge was that the IOUs, best equipped to implement these policy initiatives, did not want to participate in implementation of the policy (Basin Electric Power Cooperative, 2016g). It is critical to understand how the nonparticipation of the IOUs compelled the federal government to create an alternative to the dominant for-profit electric utilities.

From their very inception, the IOUs had worked aggressively to secure their position as the sole monopoly providers of electricity by codifying their practices under the guise of government regulatory oversight. This codification then gave the appearance of a publicly accepted "regulatory compact," which was, in reality, political gamesmanship. Samuel Insull, the figurehead of this movement and an electric utility executive presenting at a conference of IOUs, "suggested that competition was not in their best interest, and that the companies should together promote the idea of state regulation of utilities in return for the granting of monopoly service territories. Insull felt that this would ensure rapid industry growth with minimum duplication of physical plant" (Lowery, 2010, p. 5).

While government intervention into the electric power industry might seem like a ceding of power to government regulators, in reality it was a high-stakes game of chess meant to cement market dominance: "Investor-owned companies were able to establish virtual control over state regulators; and with the guaranteed income of protected monopoly service, they proceeded to build major enterprises through the use of holding companies" (Lowery, 2010, p. 6).

The bargain the IOUs made with the government paid off handsomely. IOUs were rewarded with monopoly service territories, locking in a cycle of guaranteed margins along with taxpayer-subsidized risk mitigation. Urban dwellers could benefit from all the services and conveniences electricity had to offer, and the IOUs could focus on the dense, profitable cities, turning their backs on the development of the low-density countryside whose high overhead service territory would significantly reduce their return on investment. The political calculations of the IOUs would end up creating the political backlash and animus necessary to initiate government policies addressing major urban and rural development deficits and, ironically enough, creating a populist alternative to the investor-ownership model of public utilities.

Mobilizing Community and Government Resources to Industrialize Rural America

The policy advancing public ownership of electric systems was founded against the backdrop of government support for utility monopolies, corporate influence over the political process, and wealth divides between urban and rural populations. Initially, the FDR administration sought a utilitarian approach to bridge the urban-rural industrialization divide: subsidizing the existing IOUs to extend into the rural countryside. However, the IOUs were wary of contributing to FDR's New Deal policy of absolute rural electrification, seeing the inevitability of reduced margins resulting from the addition of low-margin service territory to their profitable urban customer base, not to mention further government encroachment into their business. Indeed, an early report by the IOUs to the FDR administration sought a "no pain, all gain" policy in return for their participation (Lowery, 2010). This rejection by IOUs of incentives from the Rural Electrification Administration (REA) and the lack of corporate buy-in drove the FDR administration to create infrastructural and capital investment capacity necessary to encourage publicly owned electric utilities to emerge, to account for the failure of industry to contribute to the everyday needs of rural America.

The publicly owned electric utility concept grew out of recognition that monopoly utilities were not only falling short on their social compact but also stymieing progress toward national electrification. The broader concept grew out of a synthesis of social movements, economic stimulus policies, and prescriptions to pragmatically address core rural needs.

Electric cooperatives, in particular, had existed in the United States since the late 1800s. President Theodore Roosevelt noted the value of cooperative entrepreneurship in 1909: "The cooperative plan is the best plan of organization wherever men have the right spirit to carry it out. Under this plan any business undertaking is managed by a committee; every man has one vote and only one vote; everyone gets profits according to what he sells or buys or supplies. It develops individual responsibility and has a moral as well as financial value over any other plan" (Lowery, 2010, p. 6). Despite the explicit acknowledgment of the capacity of cooperatives to build community, electric cooperatives would not emerge as a significant organizational force until the 1930s, due in part to the backing of the federal government.

Progressive Era leaders had for years advocated for the grassroots ownership of electric systems. The Public Ownership League of America influenced many core policy solutions adopted by prominent legal figures, and its influence was profound in electrified energy. The 1920s saw Pennsylvania governor Gifford Pinchot advocate for "giant power." Power generation and transmission would be centralized at the mouth of coal mines and transported over vast expanses of transmissions lines, everywhere (the spoke-and-wheel system, in which a central generator would extend transmission lines outward over great distances, looking much like a spokes in a bicycle wheel). These visions challenged the decentralized, regionally oriented monopoly hold of energy generators that stymied access to the critical infrastructure needed to produce and transmit electricity outside of metropolitan areas (Hughes, 1976). Perhaps more radical, the expressed purpose would be not for profit but for the public good (University of Wisconsin Center for Cooperatives, 2007).

Many policies were implemented to encourage the growth of electric utilities into the countryside. Economic efficacy, more than social change, was the driving rationale behind government policy, though social movements certainly helped drive policy adoption. Those oriented toward public ownership were incentivized in basically the same manner as IOUs, though not necessarily due to their at-cost, member-service orientation, for the mere fact that REMCs were willing and able to meet the goals of rural electrification. The FDR administration grew public ownership from an idea to a reality (Greer, 2008).

Expanding access to government funding beyond IOUs to cooperatives enabled public power and through that major public works projects. Preference power—the policy giving cooperatives and municipal utilities privileged access to cheap, bountiful electricity produced from a handful of these public works projects—further developed the infrastructure and value chain necessary for the electric co-op sector to thrive. One prominent preference power public works project, the Tennessee Valley Authority, made "giant power" a reality through a demonstration of a proof of concept. Now communities looking to start electric cooperatives could count on a ready source of affordable finance and wholesale energy to distribute to their members throughout the rural countryside.

It was estimated that by the 1960s all corners of the United States were electrified either by an IOU, municipal, or cooperative. The U.S. government's 100 percent electrification policy goal was a success. The combination of readily available low-interest financing from the REA, generation and transmission capacity from preference and giant-power projects, and community organizing/public entrepreneurs in the rural countryside stimulated the growth of the electric cooperative sector. Importantly, the federal monies provided to the cooperative sector were paid back with interest, accruing a net profit to the federal government (this model would serve as a justification for the rural cooperative telecom movement as well). Just as government regulatory power was used to stabilize investor-owned electric utilities, the New Deal era of government saw a pendulum swing toward empowering public ownership options, with a net profit provided to the federal treasury.

Community-Owned Giant Power:
The Emergence of Basin Electric Power Cooperative

A small portion of the American countryside was electrified at the outset of the Great Depression. The Great Plains region was particularly neglected, with only 3.5 percent of the region electrified (Basin Electric Power Cooperative, 2016g). Power plants handled local to regional baseloads and would need significant capital outlays to supply the demands of residents in the rural regions through new system build-outs. The small-scale and decentralized nature of the existing electric power plants meant that even if the Great Plains were to get wired, utilities faced an obstacle in procuring affordable electricity and transmitting it across vast expanses of land.

Initially, distribution cooperatives were able to count on the expansion of the federal government's giant-power policy, allowing the movement to flourish:

> The construction of Missouri River Dams to make electricity and control flooding was promoted for years by President Franklin D. Roosevelt . . . and studied by the U.S. Army Corps of Engineers, but it took the flood of 1943 (estimated damage: $26 million) to convince Congress to take action and build the dams. Except for Fort Peck (a Depression Era project), the Missouri River dams were built in the mid-1950s to the mid-1960s and were the main source of power for regional distribution cooperatives. (Basin Electric Power Cooperative, 2016a, n.p.)

Policy makers made it clear that it was only a matter of time before the federal government would wind down its involvement in the expansion of preference power capacity.

The distribution cooperatives needed a way to transmit electricity over great distances while keeping consumer costs down absent of giant power. Distribution cooperatives pooled resources to form a variety of G&Ts in the region, with each G&T being owned by cooperative organizations. This allowed distribution cooperatives to secure large quantities of electricity at more predictable wholesale rates, passing the savings on to the cooperative member-ownership.

A select group of G&Ts began constructing additional electric generation capacity to break their complete dependency on federally owned electric generators. These G&Ts coordinated and managed the expansion of the cooperative-owned electric grid infrastructure. However, the capital intensiveness of such projects was cost-prohibitive to a number of the smaller G&Ts, which were left to purchase wholesale PPAs from IOUs as opposed to expanding their generation capacity (Basin Electric Power Cooperative, 2016a, 2016g).

The G&Ts of the Great Plains went through a significant strategic shift during this period. The breadth of government collaboration with the cooperatives began to draw down by the 1950s. Leland Olds, a visionary in the evolution of the U.S. electric cooperative sector, and the federal power commissioner under the FDR administration, "spoke publicly to cooperatives about abandoning the idea of building smaller generation facilities to provide power for individual G&Ts, and endorse the construction of a large or 'super G&T.' This super-G&T could build huge, coal-fired power plants that would provide power for an entire region of the country" (Basin Electric Power Cooperative, 2016a, n.p.).

The idea was a cooperative version of the federal government's giant-power policy. The purpose was for the electric cooperative system to reach such a scale that allowed cooperatives to participate in the vertical integration of industrial segments that fed into the electric grid. Once this scale is reached, cooperatives might become systematically interdependent instead of relying on the federal government while better controlling for market-based volatility (Hughes, 1976). The idea took hold: "Responding to Leland Olds' vision of a regional power supplier, East River Electric and ten other power supply systems created the Giant Power Cooperative on Oct. 4, 1960—the precursor to Basin Electric. Giant Power Cooperative developed a 10-year plan (1965–1975) for securing power supply. The plan identified two possible sites for construction of a power plant: Garrison, North Dakota and Vermillion, South Dakota" (Basin Electric Power Cooperative, 2016a, n.p.).

The governance structure of such a cooperative could be a messy affair should the plan call for the consolidation of individual G&Ts into a larger, singular super-cooperative. Organizational sovereignty would have been sacrificed for the seeming elegance of unified consolidated administration. Instead, the prospective member-owners of the new super-G&T crafted a less disruptive solution that allowed individual cooperatives to continue to exist, giving that local member base a more democratically accessible, representative super-cooperative. The existing G&Ts and distribution cooperatives would remain, and the super-G&T—Basin—would be layered on top of that system. In this manner, cooperative management and boards expressed a desire to maintain overlapping responsibilities, a decidedly "uncorporate" decision in its seeming inefficiency (whereby efficiency is measured merely by the singular bottom line of profit in the case of government-regulated utilities or clear command-control in the case of mainstream public administration).

Electric Cooperative Vertical Integration Within the Great Plains Electric Grid

The planning, build-out, and operation of Basin's first power plant serves as a remarkable example of how individuals can work collectively across multiple scales to steward complex projects among themselves (first- and second-tier electric cooperatives, as well as integrating into government- and IOU-owned electric grids). The story of Basin's first power plant demonstrates the public entrepreneurial capability of individuals outside of the perceived elite expert classes:

> [Basin], after securing the financing for a power plant, the focus became finding an adequate site to build a power plant. Adequate water supply was needed, using existing federal power lines instead of building a lot of new transmission, and securing low-cost fuel were important factors in determining the plant's site. North Dakota received immediate consideration because of its abundant lignite coal and the mine-mouth-to meter capability. "Mine mouth-to-meter" means that all elements for producing baseload electricity are in one place: the water, the coal, and WAPA's transmission system. Only 12 miles of transmission had to be built to connect LOS to the Federal power grid. (Basin Electric Power Cooperative, 2016b, n.p.)

Actors with Basin harnessed a complex sociotechnological system (comprising systems at multiple scales) meet a single goal: provide stable, affordable electricity for their member-owners.

The level of sophisticated planning has continued to evolve. Basin would build additional coal-fired power plants, peaking facilities, and transmission infrastructure.[6] The organizational emphasis on vertical integration resulted in a complete electric energy system for Great Plains residents. The website of Basin explains their operational model as follows: "Basin Electric is a generation and transmission cooperative with a three-tier delivery system: We sell wholesale power to our Class A members and others." Class A members comprise G&T electric cooperatives. "The Class A members sell power to their distribution cooperatives (Basin Electric classifies distribution cooperatives as Class 'C' members) who, in turn, sell power to retail customers. There are also special membership categories entitled Class B and Class D members" (Basin Electric Power Cooperative, 2016e, n.p.).

Basin's ownership is its member-customer base, confined to a specified service territory (however, when surplus energy is produced, Basin will sell it on the spot market, broadening the potential customer pool to other wholesale buyers, thereby increasing marginal returns). Basin's 135 member-owner cooperatives are divided into districts by regional orientation, with each district receiving a seat on Basin's board of directors. The governing board structure assigns specific regions to seats on the Basin board.[7]

Enhancing the Capacity for Collective Action

The emergence and organizational robustness of Basin demonstrate the immense capacity of individuals to work collectively to meet common ends. Here, the end—

a unified, commonly owned electric energy transmission system—is incredibly complex: electricity governance intersects with the market and state regulatory forces, and energy investment is capital intensive, demanding rigorous project planning and implementation. And when all is said and done, basic operational functionality is a technically laborious endeavor all on its own.

Key leadership expressed a need—rural electrification—and mobilized essential constituencies to procure the necessary resources. Those constituencies came from rural and agricultural sectors (the farm bureaus played a major role in parlaying their political capital; local official, July 6, 2012) to pressure the federal government for the capital necessary to kick-start the electric cooperative sector.

This process took decades and was assisted by the perfect political storm of the Great Depression and New Deal politics that brought the political capital necessary to make the vision a reality. Relying on the government or other centralized political forces may be time-consuming, cumbersome, and in the end relatively risky; that is not the point. The point here is that communities of individuals, even marginalized communities, have a great capacity to perform vitally important, complicated development initiatives. Moreover, communities may take many paths to achieve remarkable development feats, including the utilization of centralized actors such as the government. In this vein, stymieing public entrepreneurship to benefit a few (corporate investors) to the detriment of the many (rural residents) should not be viewed as optimal public policy.

That said, any institution will have its shortcomings. Even well-intentioned institutional design may result in a destructive enterprise. Institutions are crafted and shaped by fallible human beings. Robust, reflective institutional design accounts and mitigates against such fallibility (free riding, corrosive actors, and institutional capture, to name a few). Electric cooperatives are just as susceptible to these dilemmas as any other institutional form.

The member orientation of Basin, and of electric cooperatives in general, translates to a relatively conservative, risk-adverse entrepreneurial culture (Finzel & Kildegaard, 2013). Capital works projects are planned years if not decades in advance, with the fate of the cooperative's member-ownership at stake (not to mention the economic engine of the cooperative itself, in the form of jobs, tax revenue, and more). The meticulous, methodical nature of electric cooperatives is set against a core premise that has been with the sector since its inception: "Our mission is to provide the highest quality electricity at the lowest price" (cooperative official, Aug. 8, 2011). That consumer cost emphasis means that some electric cooperatives miss the proverbial forest for the trees. When a sudden shift in public political sentiment or ecological change occurs, electric cooperatives tend to be reactionary. On issues such as climate change, air and water pollution, and responsible operational governance, the sector has endured harsh scrutiny from consumer, environmental, and demutualization advocates (regulatory official, Sept. 9, 2010). External pressure from government or media is often assessed from a narrow financialized perspective and seen as a regulatory threat directed at self-governance and institutional sovereignty. One CEO of an Illinois G&T summed it up this sentiment: "Co-ops could rather do it on their own than be government regulated" (Oct. 26, 2010). The antagonism toward government is remarkable considering the role of government in jump-starting the sector.

Further compounding the dilemmas that electric cooperatives face is the so-called favorable tax status, derived in part from the not-for-profit, member-owner orientation. Cooperatives are often conferred this status because they are service rather than profit oriented. IOUs point to this tax advantage as an unfair policy privilege benefiting co-ops; this is simply a ploy to play on the ignorance of policy makers as it relates to the design of the cooperative model; since the co-op does not reward an investor class with a profit share, there is simply no profit to be taxed. But the exclusion of cooperatives from various tax liabilities means cooperatives are also excluded from a substantial government policy portfolio based on offsetting tax liability. Electric cooperatives have little to no federal tax liability, essentially isolating the entire sector from accessing advantages available to IOUs alone. Consider that virtually any electric utility project requiring financing will require significant capital infusion. As noted in Chapter Two, IOUs may leverage the federal tax benefit to procure capital investment with large financial firms as a lending incentive (if you lend to us, not only will you get a favorable rate, but you can also offset your taxes). A Basin official summed up the implications: "What co-ops gain in tax exemptions, they lose in government subsidy" (Aug. 17, 2011). In fact, the Congressional Research Service has found that the government gives greater support to the IOUs than to cooperatives (cooperative association official, Sept. 8, 2010). This government support stems from tax credits originating in the IOUs' guaranteed margins (IOUs are allowed by many state public utility commission regulators to generate margins up to 15 percent or more, whereas self-governing co-ops typically confine their margin to the 0–4 percent range).

Electric cooperatives want to be treated as different and unique, but they are embedded in a system that is inordinately influenced by the IOUs. Being embedded in an investor-dominated industry makes cooperatives more beholden to government policy developed by powerful, for-profit interests, thereby weakening the value proposition potential provided by co-ops. If electric cooperatives had sought simple solutions to complex problems, path dependency would dictate that the profit motive would have implanted itself into the system to entice the necessary investment capital. Lending costs would have skyrocketed, workforce development would have been cut, and the cost to rural consumers would not be on parity with their urban counterparts (cooperative association official, Sept. 8, 2010). So while electric cooperatives have mostly fended off these opposing forces, the sector must guard against the potential for the logical orientation of IOUs from taking home.

Cooperative advocates in government, as well as institutional actors, have mobilized many resources to address both the conservative culture and structural impediments. The pragmatic design surrounding these resources that have arisen during the twentieth century has attempted to strike a balance between respecting the core cooperative values (that paradoxically may be partially to blame for the system's conservative, reactionary tack) while promoting responsible entrepreneurship against a hostile policy and market environment.

Principal actors in the electric cooperative sector have intentionally embedded the system in a number of networks that helps provide the capital necessary for growth while also guaranteeing varying levels of successful practices are followed. Virtually all of these networks regulate electric cooperatives on the basis of participation, the idea being that "if you do not want to be regulated, then you do not have to work with us" (cooperative association official, Sept. 8, 2010). Regulation is basically voluntary. Both the federal government and cooperative organizations play overlapping, redundant roles in this, layering many governance mechanisms that build self-monitoring, reciprocity, and robustness into the sector.

The USDA's Rural Utilities Service (RUS) has for decades existed as an enabling force for growing the electric cooperative system, providing critical loans during the sector's early years. Taking out a loan with the RUS requires reporting on key performance indicators as part of the loan terms. This soft regulatory mechanism is particularly useful in those instances when electric cooperatives seek loans to stem budget shortfalls due to poor management practices. A Basin executive thought this to be a positive feature of the RUS: "RUS is like a mother. They make sure you don't go too deep" (Aug. 9, 2011). Another major electric cooperative lender, CoBank, which that provides loans for utility and agricultural cooperatives, has also used the model of soft regulation developed by the RUS. The soft regulation model continues to be an essential feature of the electric cooperative system to this day.

Systemic Resilience

A number of other actors within the system have a "parental" role similar to that of RUS. But for this research the interest is where the key actors within the electric cooperative system have been instrumental in intentionally developing system-wide robustness for their member organizations. Had electric cooperatives remained reliant solely on bank or government lending, they would have been at a severe disadvantage compared to their IOU counterparts. Influential lenders like the RUS and CoBank are not tied specifically to the electric cooperative system; they also lend to other co-op sectors and are more prone to adapt to the prevailing political winds—the RUS in particular. A number of electric cooperatives, in need of a lender tied closer to their unique needs and separated from the political fluctuations of the RUS, founded the National Rural Utilities Cooperative Finance Corporation (CFC) in the late 1960s (National Rural Utilities, 2013b). CFC has positioned itself to take the reins from government lending and enhance the electric cooperative system's ability to meet its own capital investment needs.

Investing in infrastructure is not the only means by which electric cooperatives flourish. Actors within the system have worked to enhance support services for the electric cooperative sector, to guarantee constant workforce development, strong member-owner engagement, and institutional innovation through specialized service providers. The electric cooperatives are quite aggressive at purposefully designing redundant mechanisms throughout the system that build in robustness among the organizational membership.

The National Rural Electric Cooperative Association (NRECA), the premier electric cooperative federation, is the leader in this arena, providing government relations and member services and promoting standards and best practices by which all of the member cooperatives adhere to. The NRECA can pool its member cooperatives' employee benefits monies to provide benefits packages competitive with those in the corporate and municipal sectors (National Rural Utilities, 2013a). NRECA has also created healthy retirement programs that its member cooperatives may opt into for their operational staff. (Various interviewees claimed the retirement fund is so robust that it could pay out all obligations for the next forty years, even if no one paid another cent into it.) Electric cooperatives are then able to participate in a networked system that builds in the capacity necessary to compete with the larger private-sector firms for skilled staff and labor.

The NRECA, keen to leverage collective messaging and public outreach, started Touchstone Energy, the premier marketing arm for electric cooperatives, which "provides innovative resources and the strength of a national network of co-ops, helping them enhance their unique relationships with their local member-owners. More than 710 Touchstone Energy cooperatives in 46 states deliver energy and energy solutions to more than 27 million members every day. Touchstone Energy helps its cooperatives communicate the cooperative difference to business and residential member-owners, large and small, all across the country."

Touchstone Energy, as noted on the NRECA website, reinforces the necessity of value beyond profit alone: "Four values are the foundation of every Touchstone Energy co-op's service to its members. These values represent the cooperative difference and how Touchstone Energy cooperatives connect with and earn the trust of millions of people, every day" (FreeState Electric Cooperative, 2016, n.p.). The values proposition (innovation, accountability, integrity, and commitment to community) reinforces the ICA cooperative principles and values (CPs; see Table I.1) perceived as central to the effectiveness of the cooperative model.

The NRECA also worked with a number of its G&T member cooperatives to create an energy wholesaler, the Alliance for Cooperative Energy Services Power Marketing (ACES). According to the ACES website: "Since its formation in February 1999, APM has become a nationally recognized wholesale energy trading and risk management firm that has maintained its client-oriented focus of providing quality service. Today, APM is one of the largest physical electricity traders in the nation" (ACES, 2016, n.p.).

Not only does ACES purchase wholesale power on behalf of its member-owner organizations, but it also seeks to enhance the capabilities of the staff of member cooperatives to better interact in the energy marketplace, particular regarding risk management, a critical skill for cooperatives at the mercy of a volatile energy market. This capabilities enhancement is yet another example of how cooperative support systems build resilience throughout the network.

One of the criticisms of the electric cooperative sector has been the inordinate role that coal plays in the overall energy generation portfolio. Some circles have called for the privatization (or demutualization) of those electric cooperatives slow to adapt to member needs (Cooper, 2008). That prescription does not account for the structural deficits facing cooperatives, originating in part from the long-term reliance on coal. These structural deficits have furthered dependency on coal as a dominant energy source, complicating the capability of electric cooperatives to develop renewable energy projects. NRECA was involved in the creation of another association, intended to fill a systemic need to grow the share of renewables in the portfolio of the electric cooperative sector: the National Renewables Cooperative Organization, which fills the renewable energy development gaps of many cooperatives by specializing in the arena, relieving individual electric cooperatives of the burden of spreading scarce resources thin to explore new, costly ventures.

Systems-level entrepreneurship has been a critical element of the success of the electric cooperative system. Virtually all of these aforementioned affiliated organizations require adherence to set standards, as well as reports on those ends. The electric cooperative sector is working actively on optimal fiscal governance, marketing, and community relations, enhancing the diversity of their energy source portfolio and maintaining a significant market presence. This is a critical, self-regulatory mechanism of electric cooperatives by their peers, encouraging entrepreneurship and innovation at multiple scales of the system itself. Indeed, as discussed throughout this chapter, electric cooperatives do not have the luxury of not being entrepreneurial. Fostering a strong culture of successful practices (good governance, appropriate margins, and member democratic participation) is essential for the electric cooperative system to break with historical dependencies, address substantive criticism, and meet member needs.

Path dependency, specifically regarding government privileging of IOUs for service territory and access to capital, has hampered overall adaptive capacity. Many electric cooperatives waited decades before building counter- or alternative organizations meant to enhance the system. While actors with the electric cooperative system are on a purposive path to building those necessary self-help enhancements, it may take some time for electric cooperatives to operate on a fair, level playing field with their IOU counterparts.

Considering the structural impediments to electric cooperative development of renewable energy generation, as well as the risk-adverse institutional culture of electric cooperatives, what factors propelled Basin to construct the nation's first cooperative-owned wind farm? Were pressure groups in part behind the endeavor? Did competitive activity by the IOUs play a role? Did the electric cooperative support system facilitate this project?

Sowing Seeds for the First Cooperative Wind Farm: Collective Action and Leadership in the Electric Cooperative System

Electric cooperatives covet their culture of self-regulation. Basin, and electric cooperatives in general, is rarely forced into new projects or initiatives by government mandates (like the RPS). These cooperatives are engaged in influencing government public policy, providing fierce opposition to virtually all mandatory regulation on the electric cooperative sector (regulatory official, Sept. 7, 2010). As mentioned previously, this tactic does have systemic disadvantages that in the end may cause more to harm than benefit for the sector due to the opening left to competing institutional models of electric energy governance (i.e., the IOUs) to game the system. Government-driven market incentives rarely influence electric cooperatives for the simple reason that such incentives are most often directed at market- or profit-oriented firms. Electric cooperatives have a difficult time conveying to government policy makers both the necessity of the tax exemption and the importance of putting electric cooperatives on parity with IOUs, and IOUs play on the ignorance of policy makers.

Leadership across the electric cooperative sector persistently notes that these organizations are member oriented and that members dictate change. However, the change typically comes through the formal channels of representative democracy (i.e., via elected board members and executive staff) and rarely through direct democratic mechanisms, particularly at the annual membership meetings. Indeed, some public advocacy groups have chastised what they characterize as the undemocratic nature of specific electric cooperatives. The research in this area is seemingly nonexistent. However, there is anecdotal evidence that electric cooperative actors do pay close attention to many other signals that indicate a desire for change among the owner-membership. One such signal that carries weight is the observed patterns of state regulation, pending or otherwise, on their IOU counterparts. The phenomenon of state-level RPS is particularly relevant to this case study.

Basin and Cooperative Wind Energy Development

Seven of the nine states Basin and its member cooperatives operate within have an RPS, of which five are mandatory for IOUs yet not applicable to electric cooperatives. Electric cooperative representatives from these states felt pressured to add renewable energy to their portfolio to keep parity with their IOU counterparts (association official, Aug. 9, 2011). The Minnesota member cooperatives, for example, were under intense pressure as the state's largest IOU, Xcel, is mandated to have 30 percent of its energy sourced through renewables (N.C. Clean Energy Technology Center 2012). Allowing IOUs, which already carry a great deal of political clout, to gain a greater foothold in favorable public relations from marketing their substantial investment in renewables could reinforce public perceptions of electric cooperative as out of touch or dependent on polluting technologies (association official, Aug. 19, 2011). Basin's member cooperatives are acutely aware of the potential for public relations crises and have taken steps to mitigate this.

Basin's cooperative ownership convened a meeting in 2005 in part to discuss increased ownership in renewable energy generation (Basin Electric Power Cooperative, 2009). The agenda item was a reaction to the growing list of state mandates. After Basin's decades of existence, coal and natural gas exceeded 70 percent of its energy generation portfolio (Basin Electric Power Cooperative, 2016d). Without wind energy generators, the only option left to Basin's cooperative-ownership demanding wind energy would be to purchase it either on the spot market (expensive and volatile) or via a PPA (long-term contractual obligation) with an IOU. This meant that money from the cooperative system would go into the accounts of the investor-owned system. The whole purpose of a super-G&T like Basin is to create and sustain a vertically integrated electric cooperative system; every time a distribution cooperative does business outside of that network, the integrity of the cooperative system is undermined to varying degrees by reinforcing the IOU model. Basin's cooperative-owners resolved via the formal governing processes that Basin would build and own new generation capacity to voluntarily meet state standards and stay on parity with its IOU counterparts.

Overcoming Structural Barriers to
Cooperative Wind Energy

Basin was formally obligated by its cooperative-ownership via directive by the board to build, own, and operate wind energy generators. Federal government tax incentive structures complicated the task of creating a financially viable wind farm. As noted earlier, entities chartered as not-for-profit or as cooperatives do not qualify for a broad array of state and federal incentives due to their lack of "tax appetite" or tax liability. However, unlike many other electric cooperatives, Basin had an asset to leverage to take advantage of federal subsidies, providing parity with their IOU counterparts.

Basin's asset, the Dakota Gasification Company, is a for-profit subsidiary solely owned and operated by Basin. Initially owned by the DOE as part of a settled foreclosure from a private firm, Dakota Gasification was purchased by Basin in 1988 for $85 million, significantly under market value. The facility produces natural gas from coal, serving both as a peaking plant to ensure grid reliability and as a hedge against volatile natural gas prices (Bettenhausen, 2011). A nearby mine owned by Basin sources the coal, which allows the electric cooperative to produce natural gas at lower fixed cost.

Dakota Gasification has been aggressive in capital investments directed at controlling input costs and driving down its overall production costs (the plant is one of the few facilities globally that captures emitted carbon dioxide, which is then sold to Canadian firms, which use the waste product to extract oil from the Alberta tar sands). This forward thinking has allowed the synfuel facility to produce healthy margins in an otherwise volatile natural gas marketplace. Dakota Gasification remains structured as a for-profit entity so as to have a tax appetite. This allows the facility to utilize some tax advantages reserved for production of natural gas from coal and for carbon sequestration innovation. The result is an end product with remarkable price stability; in bust times, gas prices go below market costs, and in boom times the profit margins are exceptional. The facility has served three advantageous purposes for Basin as it began the development phase of its new wind farm.

First, the facility serves as backup power for the wind farm whenever the turbines fail to generate per forecasts. The peaking function provides additional system reliability. Plus, the infrastructure can withstand sporadic utilization: "The existing conventional plants provide system reliability, and there is no cost associated with additional backup for system reliability. The only incremental costs are those associated with minute-to-minute and day-to-day operation, generally referred to as ancillary services costs" (DeMeo, 2003, n.p.). The utilization of natural gas generation and outdated power plants is a standard practice by energy generators to leverage low-margin, older generation facilities for systemic redundancy or peaking when consumer demand spikes beyond projections.

Second, Basin has a cooperative institutional and fiduciary responsibility to provide wind energy at rates competitive with its IOU counterparts. Basin engineers and planners are keen to point out that when it comes to any major capital investment, they explore all available options, including the purchase from profit-oriented firms. The government incentives available for cooperatives to develop wind energy are not on parity with those available to IOUs. Since end-consumer cost was of utmost concern, Basin was able to use its experience running a for-profit subsidiary, Dakota Gasification, and to structure its wind farm as an IOU. The only way for Basin to take advantage of government subsidies was to exhibit a tax liability. The obvious choice was for Basin to structure PrairieWinds as a for-profit, limited liability corporation. Basin would be the sole owner of the wind farm, but the subsidiary is structured as a profit-generating investment for tax purposes. When asked how this would impact the integrity of the cooperative model, a Basin official noted: "Oh, there's no concern there, whatsoever. It's all managed under the same governance structure, which is a coop. Everything housed under it has to follow those guidelines. The co-op will always have the final say."

The wind farm would have to adhere to Basin's standards. And those standards dictate that all services underneath Basin's operational umbrella be owned in common by the cooperative membership. This arrangement provides Basin with all of the tax advantages of a for-profit firm while remaining nested and governed within a cooperative organization, keeping power supplied a cost-plus basis.

Third, Basin was also able to take the taxable earnings generated by Dakota Gasification (hundreds of millions) and invest it in the wind farm. The PTC allowed Basin to maintain a sizable share of its taxable earnings and direct the retained capital as an investment in the wind farm. The PTC, coupled with the MACRS, drove down costs significantly, attracted capital from within Basin's subsidiary, lowered its tax burden, and enabled Basin to justify the construction of the nation's first cooperative wind farm (Association Official, Aug. 9, 2011).

Federated Governance, Trust, and the Principle of Subsidiarity

Operating a super-cooperative like Basin could feasibly result in organizational crisis due to many collective action dilemmas. Basin comprises over 135 member cooperatives, any one of which would act in accordance with its own organizational interest. The siting of transmission lines and generators is placed based, meaning a host community will reap some level of benefit from jobs, leases, and property tax revenues (and be at the front line of prospective conflict from various local opposition groups). Member cooperatives, acting individually or in coalitions, could organize to wield inordinate influence over any number of Basin's operational decisions. Yet it seems as though conflict is nonexistent, or at least went undetected in this research.

It would be a misrepresentation of the electric cooperative system to claim electric cooperatives operate within a hierarchy; that would imply domineering actors with authority over other cooperatives. Directives come about via collective action practices that have evolved and adapted over decades. *Polycentric governance* is an appropriate label. Each organization has a structured governance system, designating mostly clear roles and responsibilities while attempting to maximize the various specializations inherent in a given cooperative. Member cooperatives remain interdependent yet sufficiently autonomous.

Basin is capable of providing a gamut of services to the broader cooperative system. But Basin runs most efficiently when it can contract specific services out to a member cooperative, thereby reducing redundant service provision and relying on localized expertise. The entire cooperative system benefits when each cooperative is collaborative, on the one hand, but is also individually entrepreneurial, on the other. Basin could harm innovation if it were to get overly involved in its member cooperatives' functions. This principle of subsidiarity is critical for optimal operations within the electric cooperative system in that the practice promotes polycentric approaches (what one cooperative creates literally hundreds could learn from). The polycentric approach respects the autonomy of the member cooperatives and reinforces self-governing capacities by internalization of entrepreneurship.

Basin is tasked with providing wholesale energy on behalf of the cooperative member-ownership and has cultivated decades of trust by amiably meeting those ends. The experience of a North Dakota electric cooperative association official indicates that Basin will do its best to ensure various beneficial development spillovers are spread throughout its member-cooperative system and their communities. Broadly speaking, electric cooperatives have cultivated a culture respecting operational specialization and service territory (another result of the conservative, risk-adverse culture stemming from long-standing, traditional practices). Prior experience and positive outcomes mean that Basin member cooperatives have vested a great deal of trust in their super-G&T to meet sector needs.

An institutional logic seems to have developed around electric cooperatives where at-cost energy production is the primary operational mission driving outcomes, furthering adherence to bureaucratic and operational business efficiency.

This core mission helps electric cooperatives focus intently on optimizing core economic competencies. "The highest quality at the lowest cost" is the oft-repeated mantra observed across many electric cooperative interviews, at a number of levels (distribution, generation and transmission, super-G&T cooperative, and associations); the variation of interpretation depends on the level of specialization of the electric cooperative. Those electric cooperatives closer to a place-based service territory—distribution cooperatives—see providing cheap electricity, along with community and economic development services, as the cost of meeting the needs of their member-owners (individuals, families, and businesses within their service territory). This orientation contrasts with those cooperatives operating within a larger scale, with a different service orientation, such as Basin, which view cost from a much narrower vantage point: producing energy and marketing services to their owner-cooperatives. This is due to Basin's specialized responsibility of providing competitively priced wholesale services to electric cooperative institutions. Basin member cooperatives add value to the commodity and distribute the procured goods to its member-owners within a given service territory.

Cooperatives serve their members, not a bank or shareholders. Basin services direct organizational needs of representatives of electric cooperative groups, spatially situated across vast terrains. Basin's field of membership is the electric cooperatives themselves. Basin's primary concern is to then meet the needs of their cooperative member-owners, not necessarily the local communities within its service territory. It is then incumbent upon the intervening cooperative member-owners to transmit and transform those assets. The individual cooperatives then are tasked to package and distribute these assets within the identified value proposition of their member-ownership.

This dynamic helps explain the differing emphases of the cooperative organizations nested within a shared, coordinated system: the emphases need not be contentious. In actuality, the design of the system and rules in use—particularly the principle of subsidiarity—seems to enhance the development functions of electric distribution cooperatives. Basin-owned projects such as PrairieWinds are designed to capture as much value as possible for the member cooperatives while limiting risks and externalities and reducing wholesale costs of electricity. The cooperative member-owners are able to receive those optimized services by letting Basin focus on its core competencies. Basin, in essence, is a capacity builder, providing assets that enable its member cooperatives. If Basin is driven to spread development benefits widely, it must have a process justifying where to expend resources and address why one member cooperative benefits above another.

What led Basin to site PrairieWinds in the backyard of one of its member cooperatives, Verendrye Electric Cooperative (VEC)?

Why Not Minot?

Basin officials noted that North Dakota's enabling policy environment makes the state a friendly place to develop wind (cooperative official, Aug. 9, 2011). State policy, as expressed by the North Dakota Industrial Commission, views North Dakota energy reserves as a source of economic development. The commission's aggressive internal goals "of increasing North Dakota's installed capacity of wind generation to 5,000 megawatts by 2020" in addition to various tax incentive mechanisms means the state is fully behind increasing the means of resource extraction and export (North Dakota Transmission Authority, 2009, p. 2). The commission is also actively involved in extracting the region's oil and wind resources for export. The signal of political will and desire to see such projects increase within the state has played a role in limiting public opposition while gaining the attention of global energy development firms.

A Basin official, when asked why Ward County, North Dakota, was chosen as the host community of PrairieWinds, noted that the biophysical attributes of the area played a role. North Dakota is resource-rich territory for wind energy: the land is flat and expansive (Aug. 9, 2011). Ward County's elevation is the highest in North Dakota, contributing to a robust wind regime—and North Dakota as a state is ranked as having the greatest wind energy generation capacity (AWEA, 2012b). Ward County's wind regime could be readily harvested and transmitted over Basin's nearby transmission lines, reducing costly overhead. It also didn't hurt that Ward County is a little over an hour's drive from the Basin headquarters in Bismarck, North Dakota, so the wind farm would be a short drive away for key staff. But perhaps the most important factor was the perceived strength of public entrepreneurship by one of Basin's member cooperatives.

The story of PrairieWinds is the story of a cooperative partnership: Basin and VEC. Organizational leadership and entrepreneurship were cited as critical factors in locating PrairieWinds in Ward County. Basin officials knew they could count on the leadership of VEC to address myriad challenges that might arise. Indeed, many state and local officials who are heavily involved in electric cooperatives claimed VEC to be one of the most progressive, entrepreneurial electric cooperatives in the nation. VEC partners with the DOE on smart-grid technologies and participates in a pioneering hydrogen-fuel program.

VEC has a champion-type leader as its general manager (GM). He is held in high esteem by his peers: "VEC's general manager was easily one of the key factors in choosing Minot. He kept pushing that member service territory is where the wind infrastructure should go to capture and maximize the economic benefits for co-op members" (association official, Aug. 9, 2011). VEC's GM is a second-generation electric cooperative GM: his father preceded him, serving as one of VEC's founders—the GM noted: "Co-ops tend to stay in the family. It's in the blood." No doubt this long-term exposure to electric cooperatives has helped this GM see the potential capabilities of electric cooperatives. Having grown up in a household supported by a career in the sector, VEC's GM possesses a great deal of historical and logistical knowledge. VEC is then more likely to try new initiatives that might otherwise dissuade newer, more cautious electric cooperative leaders.

A third critical factor for Basin's decision to build in Ward County was a wind farm proof of concept, previously carried out by VEC. Basin speculated in the 1990s that wind would become a necessary part of its generating portfolio and wanted to be prepared to act when its member cooperatives demanded it. VEC has an aggressive track record of being an early adopter of new technologies. As wind energy became a likely Basin venture (thanks in part to favorable cost projections), VEC's GM set staff to assess Ward County's wind regime, recruiting the use of MET towers in the early 2000s to measure it. The positive results allowed Basin to justify the construction of a pilot project (the Minot Wind Project's two turbines). VEC not only procured a PPA on behalf of Basin (the MAFB became the sole long-term purchaser of the wind power) but also agreed to maintain the turbines, alleviating the excess burden from Basin (cooperative official, Aug. 15, 2011). Once constructed, the two turbines operated with no identifiable controversy, created a curious site on the landscape, and provided lease and property tax revenue in a "dying" rural township and school district (the initial two test turbines generated significant tax revenue: $50,000 in year 1, $120,000 in year 2, and $400,000 in year 3).

Developing the Nation's First Utility-Scale Cooperative Wind Farm

VEC plied its deep political and social capital within the community toward the initial project, and the wind energy development paid dividends (no doubt, the involvement of the MAFB added an element of legitimacy to the project as well, engaging a number of real and symbolic trust and reciprocity elements). The initial success of the Minot Wind Project cultivated a community perception that wind energy development would be a net gain (local resident, Dec. 8, 2011). Prior success then allowed VEC to take on riskier projects with its member-ownership's consent.

Once Ward County became the apparent location for Basin's utility-scale wind farm, the adherence to the principle of subsidiarity played a critical role in the advancement of PrairieWind. Basin partnered with VEC and capitalized on VEC's social and political capital to realize the PrairieWinds project. In fact, Basin stepped back and let VEC's GM, an actor with a great deal of local knowledge and community trust, perform a significant share of the community-organizing work. Basin came to Ward County when VEC's GM needed additional support (typically in a supplementary or contract-related manner, and never for damage control purposes).

Basin could have been seen as an external agent, seeking to extract community wealth. A constant throughout the interviews was the lack of concern expressed by actors within the two cooperative organizations that the efforts in Ward County would be perceived as a solely VEC or Basin project. The subsidiarity principle, as practiced by these two institutions, meant that (among the informants interviewed in this research) Basin and VEC were often perceived by residents as virtually the same entity. A Ward County school official noted, "Basin is as well liked as VEC" (Aug. 16, 2011). Both cooperative actors built trust and operated with the perception of integrity, mutually reinforcing the public opinion of the two cooperatives.

Going into the project, VEC's GM did indeed capitalize on the earlier success of the Minot Wind Project. The proposed addition of seventy turbines (115.5 megawatts total capacity) promised to enhance prior developmental outcomes exponentially. Community meetings were organized by VEC to spread awareness and built consent around the development of the wind farms. Frequently, the meetings were held at the local school that stood to benefit the most from the tax revenue from the new wind farm, reinforcing "things to come."

VEC went about a strategy intended to build further trust and create buy-in. A VEC executive expressed a central rationale for the adherence to transparency: "There are a number of fly by night operations muddying the picture. We didn't want to misrepresent the project or get people's hopes up" (Aug. 17, 2011). The GM set up landowner meetings, making connections with local regulatory officials to speed the process along. Basin, directed by VEC's GM, built even greater linkages among the stakeholders of the project, meeting privately and publicly with local landowners and township and city officials.

Early on, VEC reached out to local hunting, wildlife, and environmental interests—mostly hunters, birders, and Sierra Club members. The GM was aggressive in preempting local protest (no organized opposition was documented). One local birder explained succinctly how the GM had his trust early on: "The co-op management lives here, so they care more about the community" (Aug. 10, 2011).

While no names were offered up, a few interviewees claimed that some locals did not want the turbines. The claims from two interviewees were that a lot of the property applicable to wind energy development is absentee-owned farmland. The farming interests are not necessarily in alignment with the interests of residential landowners within the vicinity of the development. Farmland owners viewed wind as yet another source of revenue from the land. The explicit negative feedback from the interviewees was mainly constrained to aesthetic issues. One local farmer hosting two PrairieWinds turbines on his farmland noted that the formerly endless blue sky is now dotted with turbines, "but you get used to it."

An executive from VEC noted that one of the leading farmer proponents who leased land for two Basin turbines is now regretting his decision. The executive suggested incorporating that farmer's voice in this research. When this particular farmer was asked about his regrets, he stated, "Oh, I wouldn't go that far. If I had to do it all over again, I probably would." When this anecdote was relayed back to the VEC executive, his response was, "Well . . . I suppose he just doesn't want to start any trouble." The farmer did sign a contract with a trusted local leader and perhaps did not want to be seen by an outsider as feeling betrayed or lacking trust, particularly after the public efforts put forward by the cooperatives to secure buy-in.

Basin officials frequently hosted public educational events on potential legal and technical issues. Basin exhibited remarkable openness regarding community dialogue, getting in front of controversial topics related to wind energy development (turbine noise, shadow flicker, electromagnetism); this researcher had not observed such dialogue in analyses of four other IOU wind energy development projects. The two cooperatives used these presentations for informal assurances that were parlayed into formal assurances and agreements. A Basin official noted: "Have you seen how much up-front money we spend on impact assessments?! Well over $1 million" (Aug. 17, 2011). Clearly, the co-ops wanted to control for risk as much as possible.

Basin and VEC officials would often contrast their practices to those of the IOUs speculating on wind energy development in the region. Basin officials were quick to compare how the speculators treated local landowners (cooperative official, Aug. 17, 2011). Basin officials were proud to note they paid prospective lessors $10 an acre for a three-year speculative lease, plus a $1,000 stipend for a lawyer to review the lease terms. VEC would encourage the landowning lessors to work together, pool their money, and hire a shared legal team. In comparison, the IOUs were known for ten- to twenty-year-long speculative lease terms at $1 a year, with no stipend for legal advice. The co-ops encouraged monitoring from the community.

Basin and VEC had complete buy-in from their identified landowner base. Basin was keen to accommodate, working intently with local landowners on where to locate the roads and access points. The cooperatives noted the importance of limiting the disturbance to local landowners, expressing that extreme caution has been extended toward the maintenance and upkeep of turbine access roads and security gates. Even then, the best-laid plans do not always work out accordingly. A Basin executive noted:

You know, we sat down with all of these guys and ran a number of scenarios as to where to situate the access roads (we did not want to interfere with the crop harvest). Now we have a couple of these guys who call us up and moan about how the access roads are a pain for their harvesters to get around. I mean, they told us to build the road in the same spot they're now in a fit about! Thankfully, I can point to the minutes of these meeting and locate where they told us to build a road. (Aug. 17, 2011)

The early preparatory work paid off. Basin performed the due diligence and lined up the access agreements to the grid, procured investment monies from their subsidiary Dakota Gasification, and set VEC to work on cultivating the host community. VEC quickly satisfied the local government regulators and rapidly built up the community trust and landowner buy-in necessary to deploy the infrastructure. The long-running relationship with the community by VEC lubricated any potential social tensions, speeding the PraireWinds project along from start to finish within eight months, whereas Horizon in McLean County, Illinois, took years. Long-term trust building and strong relationships made for rapid deployment of the $250 million project (Basin Electric Power Cooperative, 2009). Such social capital will be essential for the rapid expansion of the renewable energy sector.

Direct Development Outcomes of PrairieWinds

Measuring the long-term developmental outcomes of any wind farm is difficult to do; there is no coordinated effort to track the longitudinal outcomes on local communities, and PrairieWinds is no different. That said, interviews uncovered five areas of development contributed by the PrairieWinds project: a short burst of local economic stimulus, price stability for VEC's member-owners, lease agreements, jobs, and the area property tax base.

PrairieWinds, like other wind farms, appears to have a significant upfront impact during the construction phase. The wind farm represents a $250 million investment in a relatively small, isolated area. Building tradespeople came from all over North Dakota and surrounding states. Hotels were booked up, and eateries were at capacity. As time elapsed and PrairieWinds shifted to an operational wind farm, the direct outcomes became muted. The "noise" from the oil boom diminished the perceived impacts of PrairieWinds; to many in the community, it was just another energy project. Local taxing entities and a select few property owners benefited the most from new revenue streams via lease agreements, and a handful of jobs were created.

VEC's local member-owners benefited financially, though only marginally. PrairieWinds produces electricity at a very stable, competitive rate thanks in part to the PTC, a secured PPA, and the co-op's cost-plus price orientation to the end consumer. Like many cooperatives, VEC returns excess margins or "capital credits" to its member-owners, as opposed to distributing the profits to outside investors (*Verendrye Network News*, 2009, p. 7). As consumers, the member-owners are further shielded from volatile electricity prices, and their cooperative network becomes more independent from external rent-seeking actors.

The landowners receive a guarantee of $4,000 per turbine per year. Basin secures these leases for forty years. Note that the duration of the lease agreements is longer than McLean County's Horizon wind farms, and the lease agreements are significantly less (by $6,000). Unlike Horizon, Basin did not offer good neighbor agreements, because Basin was seeking not acquiescence or to maximize profit as much as to keep its overhead costs low: "Remember, the IOU's owners are the shareholders, not the consumers. The co-op's concern is also for their owners, who are the consumers. Our goal is high quality at the lowest price for everyone" (association official, Aug. 17, 2011).

Basin executives are quick to point out that a cooperative-owned wind farm will have lower lease rates but a higher number of employees (PrairieWinds directly created eight new jobs to maintain the turbines). A corporate wind farm will attempt to keep labor costs and total employment down yet pay higher value leases to purchase community trust. To do this, these IOUs will contract out to firms that specialize in maintaining wind farms. These firms operate with a bare-bones staff and constrain services to maximize profits (wind developer, Nov. 8, 2011). So, up front, the cooperative wind farm does provide more direct employment itself but pays out less in lease agreements over more extended periods of time. The cooperative's emphasis here is on optimizing operating costs to lower overhead while enhancing service and decreasing retail costs to the its member-ownership.

Local government also benefited from the development of PrairieWinds, with Ward County receiving a modest increase in tax revenue. The rural township hosting PrairieWinds captured a great deal of new tax revenue, as well as new infrastructure built by Basin. However, it was a local school that benefited most.

The South Prairie Public School District used to be considered Ward County's "poor" K-8 school district (the principal insisted this label stuck for over fifty years). The district was experiencing a steady reduction in rural residents and a parallel decrease in property tax revenue. But the oil boom and the construction of PrairiWinds has changed all of that. The one-building school now has an enhanced revenue stream from PrairieWinds (and the new residential properties being built to house in-migrants). The district recently completed over $5 million in new renovations. The school's principal was emphatic that the wind farm has been a substantial net benefit to his district.

PrairieWinds has undoubtedly contributed to the material betterment of actors within Ward County: new jobs, new sources of income for landowners, and significantly increased revenue streams for local government agencies. However, looking at a cooperative's impact based solely on financial measures obscures the organizational outcomes. The standardized financialization of performance indicators does a disservice to the intricate patterns of activity performed by a cooperative (Borzaga & Galera, 2012). The incorporation of actor voices from the community matters for getting a sense of the community outcomes. Through this research, a number of noteworthy cooperative development activities were uncovered.

Indirect Development Outcomes: Enhancing the Capacity of Electric Cooperatives to Build Community

North Dakota has a legacy of aggressive cooperative development. During one particularly aggressive era of development led by the North Dakota Association of Rural Electric Cooperatives, in the late 1980s and early 1990s, many successful cooperatives stimulated significant economic growth. (One of those successful cooperatives, Dakota Growers Pasta, would eventually be privatized by its producer-owner farmers for over $240 million.) This era of rapid growth in cooperatives was dubbed "co-op fever" and was named the Associated Press "Story of the Year" in 1991 (Patrie, 1998).

The legacy of cooperative entrepreneurship appears culturally important. Stable, long-lasting cooperatives in North Dakota seem oriented toward more than just maximization of the bottom line: prominent North Dakotan cooperatives have a broader community orientation. Basin and VEC are two leading organizations in these regard.

Basin directly owns PrairieWinds. However, VEC, as the principle of subsidiarity, represents and is empowered by Basin within Ward County. VEC knows its community far better than Basin does, and Basin would therefore rather VEC perform community and economic development on its behalf. For analyses, it is important to realize that Basin builds capacity for VEC, and VEC converts that capacity into development and mobilization within its community. Institutional logical orientation and organizational practice matters.

These electric cooperatives are mature and always adapting. PrairieWinds—or any other major development initiative, for that matter—did not necessarily stimulate new development endeavors as much as enhance the capacity of VEC's culture of community engagement. The GM sees VEC as more than just an electric utility, and PrairieWinds enriches that value proposition.

The VEC distribution cooperative is widely respected within Ward County, throughout the region, and by its peers in the electric cooperative sector as one of the most progressive electric cooperatives in the nation (cooperative association official, Sept. 8, 2010; local resident, Aug. 8, 2011). The obvious reason that an electric cooperative like VEC garners such deep respect is that it provides tangible, material development.

Were it not for VEC's dynamism, the PrairieWinds wind farm may have been constructed elsewhere. VEC has been an aggressive early adopter of renewable energy and next-generation electric energy technology. VEC has partnered with the DOE on a hydrogen energy vehicle project, as well as smart-grid technologies, allowing it to provide enhanced cost savings to its member-owners through remote management of water heaters. VEC's entrepreneurship puts it at the forefront for testing and deploying new technologies from Basin and DOE.

Interviews and interactions with VEC's GM revealed the broad range of operational and community development activities a cooperative can perform. VEC is always looking toward new endeavors and engaged in the organizational development of its staff and member-ownership. Many colleagues in the electric co-op sector recognize the entrepreneurial acumen of VEC's GM (association official, Aug. 9, 2011). Organizational leadership infused with a vision for the community is a critical component to the development impacts of PrairieWinds and VEC. One part of that vision is the growth of new organizations to meet local needs.

As alluded to earlier, VEC was actively involved in the co-op fever era, helping new cooperatives form around the state via financial and technical assistance. The state's largest telephone cooperative, SRT Communications, was founded in the 1950s by the board of VEC. Using its extensive network to procure resources, VEC worked in nearby Berthold, North Dakota, to procure $300,000 in USDA rural development grant funds to help build a new child care center (USDA Rural Development, 2012). VEC also worked with regional cooperatives to help fund the Quentin Burdick Center for Cooperatives at North Dakota State University, which focuses mostly on agricultural issues over cooperative issues (cooperative official, Aug. 15, 2011).

VEC's influence is also felt directly in local civic activities. VEC engages and encourages its labor force to participate in local civic groups, and VEC employees are reimbursed for their membership fees in local organizations. Civic engagement endeavors also extend to the member-ownership.

VEC makes a good-faith effort to involve its member-ownership in participating in cooperative governance; transparency is critical. The website and new member-ownership orientation material come with an owner guide, detailing how member-owners can get engaged (Borzaga & Galera, 2012). In recent years, VEC has gone so far as to create a member advisory committee, open to the public. The purpose of the committee is to build new venues to engage the member-owners in an iterative dialogue that contributes to organizational policy. The committee began with twelve participants in 2007, and as of 2013 has grown to over eighty active participants, many of whom have since run for and won positions on the board. VEC's annual general membership meeting pulls 3,500 people from its base of 8,800 households. When asked how VEC achieves such significant turnout, the GM's response was simple: "A carnival and raffle for TVs and such. We make it a fun, family affair. That gets them out every time" (Aug. 15, 2011).

These features of VEC are essential. Observations of VEC demonstrate the organization's development potential. The cooperative is directly engaging their member-ownership—and, by extension, in the community—in tangible civic endeavors. The member advisory committee not only fosters a "farm team" for new board members but also elevates the member-ownership into the decision-making apparatus of the cooperative. Whereas an IOU treats its consumers as passive, VEC is cultivating actors in governance, entrepreneurship, and development or, to put it another way, democracy. VEC serves as a stabilizing force for community governance. Community wealth and decision-making authority are rooted locally. This rooting of wealth and power is of critical importance for a community enduring a series of disruptive events (the energy boom, the flood, in-migration). "Learning to craft rules that attract and encourage individuals who share norms of reciprocity and trustworthiness, or who learn them over time, is a fundamental skill needed in all democratic societies" (Ostrom, 2005, p. 133).

What the cooperative is doing appears on its face to be the necessary requisites for building civic actors and enhancing community governance. But just how empowered are the newly initiated to contribute significantly to the local social system? Is structural change possible? Or might another impediment be at play?

Co-opting Governance:
What Role Has the Growth Coalition?

Social systems are complicated. There is no doubt that VEC is building the capacity of actors in their community work collectively. But VEC is nested in many networks. Moreover, those networks determine the types of resources available, who governs over what, and the constraints and capabilities of VEC and of those that VEC seeks to empower. However, might VEC be limited to the extent to which the organization could support substantive, and possibly necessary, social structural change?

If the two case studies in this research are any indication, the vital, local groups that VEC and Basin interfaced with may be part and parcel of the standard wind energy development processes. The principal actors that regulate and interact with a completed, operational wind farm include the (a) county government (planning and zoning), (b) landowners (farm interests and residential), (c) school district, (d) township government, and (e) economic development council.

The issue here is twofold. First, the actors in this network represent a textbook growth coalition: individuals with strong incentives to prosper economic growth. Property owners and local business owners are in growth mode, seeking to capitalize on the oil boom and other development endeavors. The agricultural community sees farmland as a revenue-generating commodity. The local governmental bodies have a revenue incentive to see more property development (property tax revenue), so their values are also in line with these groups.

All interview participants in these groups noted that the Minot Area Development Corporation (MADC) is the prime mover of development endeavors in Minot. MADC appears to be a bridging or brokering organization, connecting the growth coalition interests to critical resources. Indeed, the MADC board comprises government, agricultural, financial, commercial, property, and energy interests, including the GM of VEC. Interview participants widely viewed MADC as an aggressive force for local prosperity, noting its central role in driving the oil boom (Minot Area Development Corporation, 2013a). However, one local had strong misgivings about the direction of the community under the vision of MADC: "It's mad. People don't see that they are digging their own graves. They are selling this community to outside industries that could give a damn about our livelihoods. The chickens are going to come home to roost" (Aug. 8, 2011).

Second, many of the organizations in the community's growth network have representative figures serving on VEC's board of directors (Verendrye Electric Cooperative, 2013). Some of these directors have multiple growth network connections. Six of the nine directors have agricultural interests, two of which are involved in government, and one in homebuilding.

The underlying logic of the growth network perspective is that the logic of actors within these groups tilts toward a profit orientation through the capture of the instruments of governance. It would seem to be evident that these interests indeed govern prominent local organizations interacting with the development and operation of PrairieWinds and VEC. As one farmer noted about the wind farm: "The wind isn't for ND. It's for export" (Aug. 14, 2011).

Many actors steward VEC (the staff, the GM, the board, and the member-owners), representing any number of interests. What if the growth network plays an inordinate role in stewarding the direction of VEC? There is no doubt that, as the oil boom continues unabated, the growth coalition seems to gain greater control over local resources, with growth organizations such as MADC and the chamber of commerce leading. The policies the growth coalition is pushing seem geared toward growth and bureaucratic efficiency. The principal of the South Prairie Public School District noted that, despite his district's recent resurgence and robustness, the business logic of centralizing governance permeating the town has resulted in a call for the consolidation of school districts. Out-of-town firms are being courted to develop and purchase the local property. The school district is being pressured to entice an Oregon property developer with interest in developing 500 housing units within its tax jurisdiction ("the development will bring in property tax revenue, but they want us to accommodate some of the wishes of the developer"; school official, Aug. 16, 2011). Localized ownership of business and property is diminishing, albeit with the support of the growth coalition. This unimpeded process means governance is shifting toward larger, private corporations operating from a great distance away.

There exists a sense that external actors are playing a more prominent role in local governance. This role is encouraged by the growth coalition. How do actors in the growth coalition see the role of VEC? Do they view it as just another generator of revenue for local property owners and developers? And is the growth coalition compromising or co-opting the integrity of the VEC by using it as an enabling conduit? VEC's integrity was put to the test.

The Mouse River Flood of 2011

The Mouse River Flood of the summer of 2011 was a major blow to Ward County. The flooding displaced over 10,000 people, with over 4,000 housing units destroyed. Some of the businesses that benefited most from the economic boom were reluctant to assist in disaster mitigation and recovery.

For-profit businesses were criticized for being stingy (cooperative official, Aug. 15, 2011). After some level of shaming, the local Walmart eventually contributed over $100,000 in food aid (local resident, Aug. 10, 2011). Public officials reached out to the energy companies, asking them to contribute their labor force, heavy machinery, and earth-moving equipment toward mitigating the flood damage. A local city official recounted how the oil companies refused to divert their operational staff to securing the dikes and instead cut a check for a million dollars. The problem was that the money could go only so far since the oil companies were already using the stock of heavy machinery available in the region. A local school official conveyed the thoughts of one oil executive who felt disgusted that his company thought it could buy its way out of local community obligations during the flood (Aug. 16, 2011).

VEC was a leader before, during, and after the flood. As the floodwaters began to rise, VEC set forth a number of organizational initiatives. VEC announced that once its critical facilities were secured, it would divert its heavy machinery and labor force toward mitigating the rising floodwaters. VEC provided staff with paid time to participate in building dikes and tossing sandbags despite the impact this would have on its bottom line (cooperative official, Aug. 15, 2011). The cooperative also wanted to send a clear message to its member-owners that they would not be held liable for flood-related damage to VEC's infrastructure: "The co-op has told consumer-members that late fees, facility charges, disconnection and reconnection fees related to the flooding would be waived, said Rafferty. "We know that recovering from this flood is going to take a while and we want to help our members through these difficult times" (Holly, 2011, n.p.). VEC practiced regular, open channels of communication to the broader community in order to send clear signals as to the services it would be offering and the responsibilities it would be taking on.

Once it became apparent that the floodwaters would breach the dikes, VEC began to set its sights on planning for recovery once the waters receded. VEC and Basin were instrumental in setting the philanthropic standard for disaster recovery in the area. VEC and Basin offered critical resources for those displaced by the flood. Basin voluntarily offered a plot of land to FEMA for temporary housing facilities, and VEC wired the encampment at its own cost.

VEC was able to marshal its extensive network of cooperatives to assist with disaster recovery in Minot (Holly, 2012). CoBank, SRT Communications, and VEC together pooled contributions totaling over $50,000 to the Red Cross (Cunningham, 2011). VEC reached out to its national electric cooperative network and gathered hundreds of thousands of dollars for recovery efforts (*Verendrye Network News*, 2011). When four of its employees lost their homes, VEC's board of directors approved a dollar-for-dollar match program to help, raising over $12,000 in assistance.

The flooding resulted in hundreds of millions of dollars of damage (*Minot Daily News*, 2012a). It will take years for Minot to stabilize, particularly with the escalating oil boom. Nonetheless, local officials made it clear that VEC, in partnership with Basin, set a high standard for volunteer and philanthropic recovery endeavors, despite total flood costs to VEC adding up to over $2.4 million (though VEC may qualify to have more than 90 percent of its costs covered by FEMA).

Conclusion

The operational features of VEC during the flood may serve as an extreme example of how a firm exhibiting a not-for-profit, member-owned governance structure functions when individuals within its service territory need help the most. VEC's GM and staff are persistently involved in maintaining and enhancing the community interactive features of the cooperative. Such persistent engagement reinforces broad levels of trust, a feeling of connectivity, and the capability to reach out to the member-owners and get them involved when critical issues arise. As the flood highlights, VEC is then able to harness social capital in a way that promotes stability and a sense of security.

Did the construction of PrairieWinds wind farm help develop Ward County's community structure? Not explicitly. Could VEC and Basin have done what it has (building local civics and entrepreneurship while leading the community during a disaster) without the wind farm? Chances are, yes, but this is really a longer-term question: the wind farm is both a hedge and an investment in the future. Regulation and climate change may threaten the carbon-based electric generators used by Basin. Early adoption may help Basin and its member cooperatives adapt to the future, smoothing any rough transitions and allowing member cooperatives like VEC to continue with their public entrepreneurship. Plus, by embedding electric generation capacity within Basin's nested system of owner cooperatives, the wind farms root the community wealth, preventing capital flight. Basin, VEC, and the material well-being of the community are enhanced. Locally, the construction of the wind farm did enhance the capacity of VEC. But did it also strengthen the socioeconomic grip of the growth coalition? One must look at the trajectory that has placed electric cooperatives on their current path.

Universal national electrification has been a formal policy of the U.S. government since the 1930s. The kick-start provided to electric cooperatives by the government came about due to the unwillingness of the IOUs to help meet the federal government's desired objective of 100 percent rural electrification. Electric cooperatives started out, by necessity, as agents of change and contestation against the IOUs in the governmental policy arena, but their service territories were defined and constrained by IOUs once established. Electric cooperatives, by the very act of their formation, engaged in development activities as they became permanent community staples.

As time has gone on and as residents have been born and raised with "always on" electricity, the electric cooperatives are being pressured to demonstrate a diverse value proposition beyond providing electricity (cooperative association official, Sept. 8, 2010). New in-migrants not necessarily familiar with the cooperative model take their electricity for granted ("I've always had it, so I never thought about it"; local official, July 6, 2012). Electric cooperatives are increasingly having a difficult time defining themselves as separate from the IOUs; distinguishing themselves on economic terms alone (higher pay to staff, lower executive compensation, and competitive rates) is increasingly a failing proposition. This means that the integrity of the model is not being promoted as an incubator of civic engagement, entrepreneurship, and development, a critically important point for cooperative practitioners.

Unlike a solely market-based system, the cooperative system incorporates the voices of individuals in communities generating, transmitting, and consuming energy from cooperatives. Presumably, better decisions would be made regarding siting of infrastructure, compensation for the use and consumption of resources (land, air and water), and end-consumer costs. But the strategic community economic development orientation of an electric cooperative can be diverted when influential actors, such as a growth coalition, capture the firm. This is where visionary leadership appears to matter.

The Ward County growth coalition does not seem to be controlling the direction of VEC as much as it is being harnessed by the GM to maintain and elevate local civic culture. It would not be accurate to say that VEC is actively changing the community social structure. But the engagement of member-owners in the member advisory committee and the building of new organizations is important. VEC is not developing community as much as building community capacity. It is an organization in which the mechanisms of governance remain rooted locally, waiting to be harnessed for community economic development. That is a powerful mechanism that is not part and parcel of the IOU model of governance, and as VEC's GM demonstrates, leadership is a critical component for the maximization of those assets for community ends.

The polycentric nature of the electric cooperative system and VEC's active participation is also critically important. VEC is heavily engaged in influencing public policy for its federated organizations, such as NRECA. Currently, VEC's GM is attempting to influence other cooperatives to participate actively in local community capacity building (the NRECA-sponsored Twenty-First Century Committee). This is the importance of polycentric governance and knowledge transmission; VEC, one of over nine hundred electric cooperatives, can experiment, take risks, and share its success (or failures) with other electric cooperatives, thereby stimulating new initiatives strengthening the overall cooperative system.

That said, a number of concerns related to the integrity of the cooperative model and its community orientation remain. For example, the electric cooperatives within the vicinity of the Bakken are enabling the ongoing shale oil development (Basin Electric Power Cooperative, 2012). "Basin Electric has a lot of work to do around Mountrail-Williams Electric's service territory to help meet the load growth, including building the Pioneer Generation Station, obtaining easements for the Antelope Valley Station-to-Neset 345-kilovolt transmission line, and planning substation and microwave communication site additions." Unconditional support of the growth may actually be harmful to electric cooperative member-ownership. Local public services, particularly government services, are struggling to keep pace with need. There is a critical shortage of housing, and local residents are expressing feelings of alienation. The electric cooperative could serve as a balance and apply pressure to the oil firms to contribute to enhanced community well-being. But perhaps this is extending the responsibilities of cooperatives such as VEC beyond what is reasonable. The point is that, depending on the leadership and the demands of the member-owners, VEC and other cooperatives could take on such a role; it is vitally important for cooperative stakeholders to understand the capabilities of a cooperative to enhance the livelihoods of those the institution is tasked to serve.

Regardless, it is clear that VEC and Basin are outliers not only in the wind energy sector but also among their colleagues in the electric cooperative system. It is important to understand the development differentials that may arise when diverse institutional models participate in the governance over critical collective resources. The next chapter provides comparative analysis of the two cases and explores factors that set the cooperative model apart from the investor-owned model.

CHAPTER FOUR

Comparing Investor- and Cooperative-Owned Firms

The two case studies presented in this book reveal many identifiable patterns in wind energy development processes. Structural attributes influence these patterns: government policy, market demand, public opinion, and unique local community dynamics.

The two case studies help contextualize the extent to which government policy influences the diffusion of diversified models of wind energy ownership throughout the nation. Policy has resulted in the privileging of elite actors that then fail to enhance self-determination of those communities playing host to wind energy firms. For example, it is clear that the federal government's energy policy is not as inclusive as it is made out to be.

Despite the disparity in how diverse ownership models are treated under government policy, the case studies reveal that wind energy development projects have tangible impacts on their host communities, though not always in ways that might be presumed. While both models convey benefits, impacts vary in ways that appear to correspond to the type of ownership. Development variances stemming from ownership models are somewhat nuanced until an analysis pulls back the layers obscuring finely textured details of interest (illustrated in Table 4.1). One-on-one interactions in the field helped advance understanding of these differences.

While the ownership model appears to influence community economic development outcomes, some findings from analysis of the two types of ownership appear to be relatively standard features of wind energy development. These standard aspects involve not only behavioral tendencies of those controlling the wind farm but also many structural issues that emanate primarily from the engineered embeddedness of wind energy within the electric energy industry. Each turbine is bundled in a group format, situated across vast tracts of land to maximize wind energy extraction, distributing the electricity generated to a transformer connecting to the electric grid, much like legacy generators such as coal.[1] Lead actors within the system (e.g., the FERC or ISOs) will craft much of the policy that dictates a great deal of the wind farm's operational features to ensure compatibility with and

Table 4.1. Variances in community development, by wind energy institutional model

Attribute	Corporate Model	Cooperative Model
Values	Monocentric tendencies:	Polycentric tendencies:
	External bureaucratic administration	Multilevel democratic administration
	Contractual arrangements	Community stewardship/ participation
	Proprietary information	Transparency
	Short-term, return-on-investment orientation	Long-term planning
Treatment of wealth	Wealth extraction:	Wealth rooting/redistributing:
	Negotiated community contributions	Community entrepreneurship
Treatment of costs	Externalized costs:	Internalized costs:
	Government regulated	Self-monitoring
	Subsidized financing	Mixed financing approach
	Judicial sanctioning	Community and judicial sanctioning
	Board elections	

stability of the grid at large. The firm's management and the board determine the remaining operational features. This is where ownership matters, in the stewardship of resources (constant and surplus) generated by the firm, as well as the positive externalities of the governance processes.

Electric energy actors are somewhat insulated from local public affairs. The opportunities for participatory community governance remain obscured (Florini & Sovacool, 2009, p. 5240). The operational and governance complexity of the grid is a further cognitive barrier for inclusivity and participation by community members or public entrepreneurs. A strong bureaucratic administrative element exists whereby participatory mechanisms are discouraged in exchange for expert, specialized leadership. Risk aversion, planning, and market projections are integral features of the system. These features exist in part due to an electric energy industry that is capital intensive and heavily regulated and requires a broad array of professionals in engineering, finance, law, and other relevant fields. "A mature sociotechnical system is often very conservative and its actors are unwilling to change" (Blomkvist & Larsson, 2013, p. 119). This maturation has, in turn, inculcated a culture emphasizing raw economic rationality and institutional functionalism as opposed to one explicitly concerned with social well-being or operational performance linked to civic engagement.

Discussions with regulatory officials and industry leaders depict a somewhat generalizable narrative regarding the constitutional and collective choice levels of analyses. Monocentric tendencies become a feature, and many actors engaged in electric energy demonstrate isomorphic traits, meaning that they increasingly act like one another; the lack of differentiation has an observable effect of stifling innovation. In this narrative, the IOUs have inordinate influence over the entire system, harnessing their structural advantage, which emanates predominantly through government energy policy—strengthened by their participation in the electoral politics and contributions to political candidates—predetermining standards by which capital is accessed, exacerbating structural disadvantages felt by cooperative business interests.

The build-out of new renewables and smart-grid technologies means that the grid will grow even more complicated, and with that complexity come even more barriers. Rural communities will increasingly play host to this complex generation infrastructure. However, unless government policy changes or the electric cooperative system (arguably the most relevant advocate for publicly owned and governed power) takes a more aggressive stance, communities will be in the back seat while IOUs drive the terms and conditions for development. The structural impediments place electric cooperative firms in disparity with their IOU counterparts, which in turn impacts their capacity to deliver on their value-added proposition. This observation is noteworthy considering that so much of the electric energy sector falls under some form of public ownership. That said, there is much the electric cooperatives could do.

Regardless of the ownership model, the manner with which the local community engages with the firm, as well as the actors stewarding the firm's resources, also has implications for the electric energy system itself. To compare the two case studies, chapter provides a synthesized discussion of the three core research questions:

- How do the multilayered governance systems in the United States influence local-level wind energy development (government and market policy)?
- How does wind energy development influence and interact with local community social structures (community)?
- How does the ownership model of the wind energy firm affect community development (the firm)?

The discussion assesses the implications of the renewable energy sector's orientation on community economic development, the policies reinforcing this, and the broader implications of public benefit from renewable energy development.

How Do the Multilayered U.S. Governance Systems Influence Local-Level Wind Energy Development?

The two case studies uncovered a great deal about how wind energy and the electric grid are affected by government and market policy. Governance over wind energy occurs at many levels through overlapping mechanisms by a broad array of actors. Interviews with actors in the regulatory, development, investment, and advocacy fields reveal many public and private actors, structural impediments resulting in limited wind energy ownership models (and public participation), and a broad array of motivations driving wind energy development. These revelations brought to the fore many social dilemmas facing the entire wind energy sector, as well as specific issues related to features of the ownership model of the given firm.

This section gives a brief perspective on the cooperative and IOU sectors in the electrical industry. The perspective is linked to an analysis of government energy and financial policy as it pertains to wind energy development. The analysis here is critical, because key players in this arena react to government policy, even though, as the analyses here reveal, level jumping or crosscutting among various policy processes is performed by the IOUs—directed explicitly at governmental policy makers—to shape the very system they then react to. This is not meant to intentionally put the cooperative system at a disadvantage as much as to guarantee rents for the IOUs (in the form of steady subsidies, guaranteed tax breaks that cut the cost of loans from major financiers) and to substantially reduce business risks, guaranteeing a solid return on investment.

The Electric System

Path dependency is built into the overall electric energy system from its inception, putting electric cooperatives at a structural disadvantage relative to their IOU counterparts. From the beginning, IOUs staked claims to high-margin metropolitan areas, leaving the rural countryside to fend for itself. Rural actors mobilized to address the electrification deficit. The early development of the electric system has today left cooperatives with low-density, high-overhead, low-margin infrastructure relative to the more profitable IOUs. There is far less room for error on the part of electric cooperatives, with less wiggle room on the balance sheet than the IOUs. To better control for their slim margins, electric cooperatives work extremely hard to contain their costs (costs going above and beyond basic infrastructure). Despite this, electric cooperatives remain reliant on IOUs and public power for over 50 percent of their wholesale energy needs.

IOUs achieve higher returns on their fixed-cost investment due to the higher density of consumers within their service territories (more households, less infrastructure). The electric cooperatives have a long-standing orientation of operating under significantly smaller margins since they have higher overhead and limited means for spreading the cost burden (fewer households, more infrastructure). One electric cooperative official claims that the at-cost orientation of electric cooperatives has made them so robust that "if we [electric cooperatives] were competing over the same utility lines, no doubt we would beat them on price" (Sept. 8, 2010). Policy and the public interest would be better served by understanding how co-ops might deliver value in denser, urban territories, and IOUs in sprawled rural areas.

Despite the current disparity, it is difficult to say if the electric cooperative system would have penetrated this far into the electricity sector had it not been for the intervention of the U.S. government. The FDR administration's creation of the Rural Electrification Administration (REA), with its recalibrated policy treating electric cooperatives as equal to IOUs regarding eligibility for government assistance and its support system of professionals, was instrumental in breaking the absolute stranglehold of IOUs. Additionally, the REA drafted a model state law (the Electric Cooperative Corporation Act) that further assisted state and local efforts to rapidly advance the early build-out of the electric cooperative system. The adoption of model statutes was vital in helping to create a more competitive environment against the monopolistic tendencies of the established IOU system, nurturing institutional diversity on competing visions for delivering public goods, and extending electric energy to low-margin rural regions of the nation. The electric cooperative system has helped balance the tendencies of the IOU system, thereby imparting a stabilizing effect to the electric energy commons.

The FDR administration's purpose for empowering cooperatives was due in part to their core design: the cooperative model is meant to grow, innovate, and catalyze approaches for self-help development. In this manner, cooperatives are not wholly without recourse, as is evident in the case of Basin Electric Power Cooperative. The study of Basin, and its cooperative member-owner VEC, reveals that cooperatives have created sophisticated federated governance systems. These systems are polycentric, with multiple levels of decision making and responsibility and an emphasis on building entrepreneurial capacities and enhancing robustness within the member networks and organizations. The creation of associations at the state, regional, and national levels has formalized these structures. The nesting of cooperatives into extensive associations of cooperatives allows for dynamic economies of scale and the enhanced provision of goods and services. Nesting then strengthens the capacity of locally rooted electric cooperatives to concentrate their member orientation and of Basin to focus its specialized service portfolio.

Again, many policy mechanisms subsidize the energy sector as a whole (see Chapter Three). A growing body of literature has documented the externalized costs of carbon-based fuels on public health and the environment. The grid itself is sustained through DOE grants and managed by federal government corporations and partnerships, enabling ongoing development initiatives by the private sector. Moreover, these energy sources are directly subsidized despite their profitability. The federal assistance provided to electric cooperatives and IOUs demonstrates the structural disparities embedded in the system: cooperatives pay interest on their loans; whereas IOUs have access to a number of additional subsidy mechanisms beyond loans, externalizing their cost of money to U.S. taxpayers. From a straight financial point of view, electric co-ops provide a net profit to the federal government; IOUs are a net loss.

A criticism of the long-standing practice of extending electrification through government subsidy mechanisms is that it has built in path dependency through policy inheritance. A report funded by the NRECA concluded: "All electric utilities in the United States receive federal assistance, or subsidies. This was the conclusion of Nobel Laureate economics professor Lawrence R. Klein of the University of Pennsylvania and has been further substantiated by numerous studies by federal agencies and others" (NRECA Twenty-First Century Cooperative Committee, 2013, p. 121).

From an operational perspective, both of these ownership models have diminished their structural capability to self-finance capital-intensive projects. Investor-owned corporations are averse to withholding significant capital reserves adequate to purchase wind farms outright (margins are distributed to shareholders as dividends and to executives as compensation). Many wind energy firms truly are extensions of venture capital, diminishing the sector's overall capacity to use retained margins for growth. Executives are motivated to return profits quickly to the investor shareholders, not to grow the total share of renewables to reduce carbon emissions.

The member-ownership of a cooperative desires a different return on investment. Cooperatives are sensitive to appearances that they are charging their member-owners excessive rates to enhance their margins. Substantive capital reserves may give the appearance that member-owners are being overcharged and that the management has ulterior motives for the money. This results in the distribution of margins back to the member-owners based on the share of patronage to demonstrate fiduciary stewardship. Short-term member-owner gratification holds up self-financing capacity building as it pertains to investments in renewables.

The ramifications for the electric cooperative system are important. The limited capacity has the effect of reducing systemic self-governing capacities outside of government-refereed actions. When the federal government provides substantial grants or backs privately funded loans, it does so with the power of the full faith

Table 4.2. Electric system comparison: rural electric cooperatives (RECs) versus investor-owned utilities (IOUs)

Market Share	RECs	IOUs	Industry Total	Source
Sales (megawatt-hours)	413,444,337	2,683,032,433	3,716,887,008	2011 EIA data
Market share	11.1%	72.2%	83.3%	
Customers	18,598,584	103,820,619	143,538,608	2011 EIA data
Market share	13.0%	72.3%	85.3%	
Distribution line miles	2,474,326	2,850,109	5,940,810	2009 RUS and Electrical World Directory of Power Producers
Market share	41.6%	48.0%	89.6%	
Customers per distribution line mile	8	36	24	
Revenue per distribution line mile	$7,000	$59,000		Basin Electric Power Cooperative, 2016f
Transmission line miles	16,005	199,924	273,564	http://www.eia.doe.gov/cneaf/electricity/chg_stru_update/chapter3.html#trans
Market share	5.9%	73.1%	79%	
Net generation (megawatt-hours)	212,749,574	3,177,369,099	4,084,681,751	2011 EIA data
Market share	5%	78%	83%	

Abbreviations: EIA, U.S. Energy Information Administration; RUS, USDA Rural Utilities Service.

and credit of the United States. Reliance on government subsidy has in turn had an impact on the organizational capacity to self-finance critical infrastructural investments. New deployments of innovative technologies are put on hold until government policy incentivizes investors by reducing or eliminating risk or enhancing profitability. Investing in energy development becomes a safe hedge for large financiers looking to guarantee a return on investment, with investment piggybacking on government subsidy. Decision making breaks down, authority becomes hierarchical, and "spontaneous" or self-led entrepreneurship is severely diminished (Aligica & Tarko, 2012, p. 258).

The total result is an energy sector reliant on large financial partnerships to make up for the lack of internally available investment capital. IOUs have a history of strong relationships with major financial partners. The relationship carries over to the operational, organizational aspects as well. A Horizon developer noted that wind energy firms commonly hire individuals from the ranks of the financial sector with experience procuring capital for development (Nov. 9, 2011). According to an Illinois regulatory official, this practice is so common that "the folks running wind energy companies are the same folks involved in oil, natural gas, and banking" (Oct. 25, 2010).

Government serves the role of greasing the squeaky wheel. And with the increasingly gridlocked government, there is a greater likelihood that wind energy development will endure many shocks, harming the deployment of electric energy infrastructure. How does the wind energy system operation in this environment?

The Wind Energy System

Growing the wind energy sector will continue to require significant, prolonged infrastructural investment. This necessity is due in part to conversion from the traditional grid model to the smart grid. The enhancement of energy reliability and independence from nondomestic sources is a federal government policy objective. A DOE official noted: "We will spend more money going from coal to wind, from wheel-to-spoke to decentralized" (Sept. 7, 2010). However, the electric cooperatives are feeling a level of strain regarding the lack of subsidy parity between the two institutional forms.

Electric cooperatives are engaged in wind energy governance but not as owners; their role is primarily as a wholesale consumer through PPAs with IOUs, which own the bulk of the infrastructure: they own almost half of all distribution lines, control over half of the transmission lines, and market to a significant majority of American consumers (see Table 4.2). Quasi-governmental organizations serve as referees of sorts, regulating the initial connection of the wind energy generator to the grid and the base load projections needed to meet demands, and as a check against price fixing. This structure challenges the electric co-ops where the system is organized by regulators who privilege IOU interests above those of co-ops.

All of the users of the grid will need to be involved in upgrading the infrastructure for wind and other renewables to grow at the pace necessary to mitigate climate change. A leader in the electric cooperative sector highlights some of the problems of shared governance in the sector: "Whose responsibility is it to modify the grid? Who does it fall upon to finance this? If the IOUs aren't throwing any of their money at it, why should cooperatives?" (co-op association official, Sept. 8, 2010). This collective action challenge underlies the difficult discussions the electricity sector is currently facing. The polycentric traits of the system undermine grid resilience due to unclear responsibilities and obligations. The issue of access to infrastructure—or rights—seems well established, but responsibility is not. The frequently voiced narrative was the need for a federal government energy policy to grow wind energy and to identify who is responsible for procuring the resources necessary to avoid an underfunded, ill-equipped grid—a tragedy of the commons. Government policy obscures responsibility and places all electric industry actors in a position of mediocre policy dependence.

Until additional efficiencies deriving from technological innovation and economies of scale make wind costs and wholesale prices competitive against subsidized hydrocarbons, or industry convenes to form a self-governing compact, government energy policy will remain a critically important component for deploying wind energy. Even if the cost of wind energy becomes price competitive, state government energy policy will probably be required to coerce utilities to make wind a significant share of their energy generation portfolio, lest capital reserves be distributed back to shareholders.

The critical role that state governments play in wind energy governance is in creating demand to stimulate growth in the sector, stimulated by state-by-state renewable portfolio standards (RPSs). The Illinois RPS mandates that electricity distributors procure 25 percent of their electricity from wind resources. Notably, one state's RPS can stimulate interstate trade in wind energy. North Dakota's RPS is voluntary and toothless, but the state's favorable development policies make it an attractive region for developers to build wind energy generators to provide electricity to Minnesota, a neighboring state demanding wind energy thanks in part to its RPS.

Analysts must also understand the calculus of the electric utilities to better assess why they function in a given way. Energy generators are not without risk: new generators are costly and always face some level of trade-off (wind's trade-off is in the intermittency and wildlife regulatory issues).[2] Wind energy, like all other forms of electric energy generation, requires enormous capital outlays. Wind, as a free fuel source, helps bring about market stability in that long-term costs are more easily accounted for than the more volatile coal or natural gas supplies (which promise to get more volatile as awareness of the linkages between greenhouse gasses and fossil fuels becomes an explicit matter of public policy). The trade-off in policy is that early-adopter costs are externalized via government subsidy to kick-start the sector.[3] Such a long-term government subsidy policy enables the development of better technologies and cheaper wind energy with reduced environmental impact as broader adoption takes hold.[4]

The most crucial subsidy mechanism is the production tax credit (PTC). Wind energy developers are strongly dependent on the PTC to offset their costs and raise investment capital by sheltering an investment partner's tax liability. The PTC makes wind energy attractive to large financial and corporate firms, particularly those looking to shelter taxes through monetization of the PTC. Wind energy IOUs are blurring the lines between energy generators and shell companies for these financial interests. Take, for example, Horizon, which changed hands many times from their initial incorporation status to Invenergy and finally to EDP Renewables. The trend looks likely to continue: a DOE official noted, "It's increasingly becoming common for IOUs to grow wind farms through unregulated subsidiaries" (Sept. 7, 2010). Critics view the PTC as a large handout that can offset more than half of the total costs of a wind farm (association official, Aug. 9, 2011). No doubt, this makes wind energy development quite lucrative: one wind energy developer interviewed in this project claimed that financiers could see debt payoffs in as little as three to seven years (Mar. 25, 2010). Large corporations with significant tax burdens are then able to reduce their burdens and profit from that process, while low-income communities are left out in the cold, paying their full share of taxes without the option of profiting from relief of their tax burden through investing in wind energy.

The government subsidy directed at wind energy follows a century-long pattern of government assistance for large-scale energy providers (see Chapter Two). With wind, the government subsidy is skewed toward the benefit of the IOUs. Electric cooperative representatives interviewed in this research were quick to point out that virtually all incentives offered to electric cooperatives had been offered to IOUs years or decades before them. Such an assertion certainly merits further study. Cooperatives are reacting to wind energy growth and development but are not deeply involved in making relevant policy; indeed, they seem more interested in preserving their legacy generators than expanding into renewables. The fault is not wholly on the IOUs; the co-ops bear some degree of blame. Despite evidence of a delay in transition and system barriers, some electric cooperatives are finding clever approaches to overcome these hurdles.

Thriving Despite the Systemic Disadvantage and Increasing Vulnerability

Government policy certainly plays a role in growing new wind energy. But these policies provide a vehicle to shirk tax liabilities by the same financial entities that shook the global economy in 2008. What are the implications of an unstable financial institutional culture being transmitted to the nascent, necessary wind energy industry? If the buzzwords *energy security* and *energy independence* are to have any meaning, shouldn't policy be aligned toward those institutional arrangements with similar value orientations? In this way, we may be privileging the firm most likely to crumble during times of volatility.

Cooperatives offer many advantages regarding sustainable institutional development. Initial research on this question has shown that cooperatives are safer investments with lower defaults than their corporate counterparts (Murray, 2011; Stringham & Lee, 2011), meaning the longevity is more dependable, critical for long-term policy planning. Government energy policy has a perverse effect in that it privileges for-profit firms through tax-offsetting subsidies. Cooperatives have historically been at a disadvantage due to their low-tax, not-for-profit structure tied to their member-owner governance, rendering them ineligible for those government subsidies. Cooperatives have limited options outside of their sector and occasional access to government assistance. One example is in the critical partnerships a cooperative needs with large financiers.

One prominent electric cooperative developer claimed that IOUs exploit the subsidy to enhance investor relationships (Aug. 15, 2011). According to his claim, backed up by many other officials within the electric cooperative sector, return on investment on IOU initiatives yields a range of 9–11 percent (the yields are presumably artificially inflated due in part to tax savings, subsidies, and overpriced end-consumer utility bills), whereas cooperatives yield a return of about 3–5 percent (if this holds true, cooperatives are better able to do more with less overhead). Cooperatives require patient investment capital, but the risk appears too high. The fact that individual distribution cooperatives are often too small to entice significant investment capital compounds this scenario. This patient capital requirement puts them on financial disparity with IOUs (though data verifying this claim does not appear readily available, it merits further study).

Despite their disadvantage, electric cooperative leaders consistently expressed their sentiment that government policy was neither out of favoritism nor rejection of the cooperative model but a fundamental misunderstanding of its owner-governance structure and how its resources are allocated. The claim is that policy makers and financiers have been trained to understand more dominant forms (government, corporations, and nonprofits). According to this persistent narrative, models like cooperatives are viewed as alternatives, going untaught in major public educational institutions. The confusion results in a general perspective that cooperatives are foreign, confusing, and seen as risk-prone due to the member-owner nature. A cooperative isn't always "legible" (Scott, 1999) or understandable to traditional finance. The member-owners of a cooperative are unlike the shareholders-owners of an IOU. A question often arises as to who is responsible for defaulting on loan repayments. Despite the fact that cooperatives share the same limited liability attributes of investor-owned firms—and therefore the same limited liability protections afforded to their ownership—cooperatives must build strong relationships with investment firms, utilize specialized banks (e.g., National Cooperative Bank), or self-finance. The illegibility to investment interests, coupled with systemic impediments, limits growth capacity of cooperatively owned wind.

The cooperative member-owners of Basin and other G&T electric cooperatives task them with attracting investment capital to assist the systemic growth of the members. NRECA has been a leader in growing access to capital for distribution cooperatives with resource deficiencies that are not members of a G&T (discussed in Chapter Three). Basin can harness the aggregate system in a manner that one distribution cooperative could never do. Plus, Basin can better justify withholding margins as capital reserves since their core mission is in part to serve the needs of their owner-cooperatives.

Basin was observed doing two things to counter the structural disadvantage regarding wind energy development:

- Basin sidestepped the not-for-profit "cooperative problem" by incorporating its wind farms as a for-profit, investor-owned entity (the sole investor being Basin), thereby becoming eligible for government subsidy.
- Basin self-financed from one of its other taxable subsidiaries, thereby negating the need to procure investment capital by interacting with large financiers, which typically do not understand the differential features of the cooperative model.

Basin was then able to exploit all of the same government policy benefits as Horizon and other IOUs while growing its service portfolio under the cooperative ownership and governance model. External ownership did not extract wealth, as occurs with IOU relationships.

Replication of the Basin approach to developing PrairieWinds by another cooperative is unlikely. Basin is one of two so-called super-G&Ts in the United States; not many electric cooperatives can leverage capital and subsidy in the manner Basin has. Officials within Basin were keen to note that it corporate subsidiary came about through opportunities and relationships with the DOE, not out of a grand design to optimize organizational agility ("the stars aligned for us"; association official, Aug. 9, 2011).

Such a development tactic opens electric cooperatives to attacks on the integrity of the model from many corners. The formation of investor-owned subsidiaries may give the appearance of money laundering (as one electric cooperative in Atlanta has been accused of doing; Cooper, 2008). The corporate entity must seek to make a taxable monetary profit, which comes from the PPA owned by Basin. Officials with Basin defended this tactic, noting that organizational wholesale costs and prices to the owner cooperatives are virtually always the first concern and that PrairieWinds was developed to provide the best cost-competitive wind power possible. Any profits are returned to Basin and administered the same way as other marginal returns in the cooperative: investing in infrastructure and returning margins to owners. Furthermore, Basin noted that its reports are in the public realm, meaning member-owners can self-monitor.

The analyses of Basin reveal an organizational culture where the G&T attempts to control for as many variables as possible. The electric cooperative system as a whole is culturally oriented toward vertical integration to maximize control for member benefit (cooperative association official, Sept. 8, 2010). The desire to control for the unknown is particularly evident when variables may interfere with its capacity to meet the service demands of its member-owners, or reliance on other types of firms makes the G&T more susceptible to price volatility.

Nonetheless, it stands that Basin and VEC have to work within rules that, from a development perspective, are designed to accommodate IOUs. Invariably, many electric cooperatives must interact with their IOU counterparts, whether it is to purchase wholesale energy, share transmission lines, or vie for service territory. None of the ownership types connecting to the grid can isolate under the current technological regime.

The origins of the IOU and cooperative systems have placed both on path dependency. IOUs are widely understood and are better positioned to leverage legibility, revenue streams, and valuation to procure capital and to quickly expand their generation and transmission capacity. Path dependency seems to play a significant role regarding the question of why is there only one cooperatively owned wind farm. Cooperatives by and large began disadvantaged: they arose to develop the low-margin service territory the IOUs rejected due to profit projections. Compounding the problem is an energy policy acting as a public subsidy for the private sector to invest in and govern over the resource-rich areas of the grid while leaving cooperatives with sparse, high-overhead rural regions. The features of the service territory then exacerbate the government policy disadvantage faced by cooperatives. Concurrent to that, cooperatives have less equity due to their decentralized structure. Plus, instead of dipping into their capital reserves or going into debt to build their own wind farm, cooperatives tend to purchase PPAs exclusively through the IOUs (since IOUs are the dominant form of wind energy ownership). "The main point though is that co-ops don't really own much wind per se. They have many more PPAs where they've committed to purchasing the power generated by a wind farm owned and operated by a private entity that built it and probably got tax advantages for doing so" (cooperative official, Aug. 11, 2011).

Government energy policy is bent in favor of investor-owned, for-profit models of wind energy production. Electric cooperatives taking a reactive stance to government policy fail to differentiate themselves on critical elements of the cooperative model and the potential benefits of cooperatives toward advancing energy policy. Cooperatives then are simply are not at the policy-making table. Public policy thus creates a chilling effect, discouraging robust collective action by a majority of U.S. energy consumers (the public service paradox, in which consumers are not given voice in the provision and procurement of a public good).

These now inherent structural deficits, coupled with IOU-centric government energy and fiscal policies, place the electric cooperative sector at a disadvantage. This disadvantage has catalyzed entrepreneurial responses by the electric cooperative sector (the formation of G&Ts, purchasing cooperatives, financial co-ops, and representative associations). Under current policy arrangements, the IOUs would appear to receive an inordinate amount of government privilege, allowing IOU models of electric generators to deploy new renewable energy generators. There is still time for cooperatives to mobilize to change these ongoing dependencies.

How Does Wind Energy Development Influence and Interact With Local Community Social Structures?

The previous section addresses the issue of policy, structural pathways, and the resultant features of wind energy development. What are the implications at the local community or host level? How are those wind energy host communities affected by this policy? Also, what might happen to these firms and their communities if government support was more evenly allocated, shifted from IOU to co-op, or removed entirely? The two case studies revealed many common features of wind energy development observable within a host community.

Three distinct phases are related to wind energy development, during which actors have varying degrees of influence and governance: (a) exploration and development, (b) construction, and (c) operations and maintenance. A wind farm, physically and socially, will have greater opportunity for involvement during the development and construction phases when the firm needs broad-based community buy-in and consent. However, once a wind farm is constructed and operational, there is little a community can do to alter how the firm distributes newly generated wealth or how it otherwise interacts with the community. Strategically, the community needs to be engaged from the start to maximize community benefit.

This section addresses the extent to which the features of wind farms have relatively generalizable interactions at the community level. This generalizability is due in part to the aforementioned government regulatory apparatus, the rigidity of the market, and the organizational design of wind farms (addressed further below).

As noted previous chapters, a community may or may not provide the social system through which its members' needs are met. It may or may not provide a sense of identity for its members. What a community does provide is what some sociologists now call *locality*, a geographically defined place where people interact. The ways that people interact shape the structures and institutions of the locality. Those structures and institutions in turn shape the activities of the people who interact. Development of community is a practice-based profession and an academic discipline that promotes participative democracy, sustainable development, rights, economic opportunity, equality, and social justice, through the organization, education, and empowerment of people within their communities—defined by locality, identity, or interest—in urban and rural settings.

Wind energy development offers many mechanisms that can strengthen the community. Several of the mechanisms have strong linkage to the features of a civically oriented polycentric system, particularly when actors are allowed to engage in the process. It no doubt takes inventiveness from the firm to ascertain the optimal linkages to community governance.

Process matters a great deal regarding community development outcomes. How people interact, and how those interactions facilitate transference of trust, knowledge, and other resources are crucial to better understanding both the impediments to community development, as well as how to better develop the community field; the flatter or more accessible the governance institutions, the greater the potential for individual and community-wide collective action through interaction (more recent work finds the social order itself is maintained when the gap in inequality is lessened, for example) (Ostrom, 2005). Community inclusivity allows for enhanced information flows toward the wind developer so it can better adapt to social dilemmas. Community inclusivity will be of critical importance throughout the remainder of this analysis.

Scouting for Ideal Communities:
Prerequisites for Wind Energy Development

Both communities in the case studies had ideal wind regimes: contiguous plots of land, ready access to the grid, and robust wind. Developing wind energy in Illinois occurred in a relative rush compared to North Dakota, for many reasons. The existing grid was built out to accommodate future growth, making a long-term investment more attractive in Illinois than in sparsely populated states like North Dakota. For example, McLean County, Illinois, has access to significant transmission infrastructure connecting to Chicago, Indianapolis, and St. Louis regional energy consumers; developers could readily plug in to the Illinois grid (association official, Nov. 3, 2010). These dense metropolitan populations are more likely to exhibit higher consumer demand for renewable electricity.

North Dakota's total wind regime is not as attractive: it is isolated from significant transmission infrastructure, requiring additional build-out of transmission lines, limiting access to the broader electricity market. The grid is constructed to service relatively rural regions, meaning the grid will need upgrading if the North Dakota wind were to be used for regional energy needs. Moreover, cooperative ownership of the transmission grid means that Basin is disincentivized from allowing for-profit firms to gobble up transmission capacity (Basin's orientation is toward its owner cooperatives, not to help provide strategic access to markets for private partners).

Once wind energy developers have determined the economic environment to be favorable, they begin the search for ideal communities nested within these environs. The attributes of a community and critical actors are significant determinants in wind energy siting. What set these two communities apart from other communities with similar wind regimes and made them ideal for wind energy development was the reception by communities. The growth coalition, with its heavy orientation toward land development (real estate, agriculture, or otherwise), smoothed the path for wind energy developers to make a case for wind energy development. That said, these firms are not unprepared for any potential hiccups along with way.

The wind energy development groups are learning a great deal from one another about how best to develop new wind farms. The information gleaned in the field is transmitted to associations and shared with colleagues and at conferences (e.g., the Center for Renewable Energy at Illinois State University in McLean County facilitates many of those discussions for Midwest-based developers; university official, May 10, 2010). A dominant theme coming from business-to-business sharing of best practices is that the greatest ally of the wind energy developer in averting organized opposition is transparency and trust building (wind developer, Nov. 8, 2011).

Wind energy developers are fully aware of the range of options opponents may use against them. Actors living within and outside of the proximity of the turbines have any number of reasons to oppose a wind farm and have some venues in the government (primarily county zoning boards) and public (e.g., local media outlets) to stop wind energy development. This study found that quality-of-life issues (including aesthetics, sense of place and social identity, cultural and rural heritage, property values, and public health and safety) and the threat to wildlife conservation efforts were reasons for opposition. A wind farm changes the skyline of a community for a generation or more. The general public may also have preconceptions as to the public health outcomes of wind energy development (turbine shadows, noise output, and harm to wildlife). The wind energy developers are keenly aware of this and proactively prepare to mitigate the numerous levers that may be used to stymie the ambitions of a wind energy firm.

Wind energy developers work diligently to foster community buy-in, adhering to an early-and-often principle of outreach. Public events are held in symbolically important venues (e.g., schools and firehouses), advancing the transparency principle, fostering greater buy-in through interaction, camaraderie, and trust generation. Transparency also applies to upfront discussions of controversies, traditionally associated with wind energy development. Additionally, wind energy developers go into each situation knowing that significant financial investment will be needed to proactively curtail environmental concerns. The wind energy firms have discovered it is best to pay for environmental impact studies up front as insurance against future criticism. While the cost of impact studies (hundreds of thousands to millions of dollars) is sizable, it is a small price for developers to pay in an industry with the promise of healthy, guaranteed margins (association official, Aug. 17, 2011).

A looming problem for wind energy development will be the inequitable distribution of project burdens and benefits, distrust in the developer's motives, and perceived lack of public participation in the decision-making process (McGinnis & Brink, 2012). Thus far, there has been little sustained opposition on this front. Perhaps this is due in part to how the wind energy developers advocate the economic impacts of a wind farm.

The developers are very keen to begin the conversation with a discussion of the financial returns of prospective wind turbine hosts. When harnessed, the wind farm can now convert an unused resource into a harnessed commodity. Buy-in was sought through appeals to individualism (you could make a pretty penny if you host one of our turbines!) and distributive communitarianism (think of the benefit to the schools! think of the children!). The appeal to individualism promised a virtually cost-free return on investment: the quarter of an acre of land is operated and maintained by the wind energy firm, with checks sent on a timely basis to the property owner. Fairness in contractual agreements was hammered home through an open process as well: "You always run into those types that think one neighbor is getting a better deal than the next. You want to nix that up front. In all of our presentations, we talk about the blanket offer for each turbine. No exceptions. If it so happens that one property owner gets an increase in lease payments, then all property owners get an increase. This is featured in all of our public meetings" (wind developer, Mar. 25, 2010).

The developers also appeal to communitarian ideals to advance their development ends. Common themes were as follows:

- Your neighbors are going to be financially better off. Property owners will make more money on their land, and we will create new, high-paying jobs.
- Your local government will get a new source of tax revenue (sparing you of local tax increases for the foreseeable future).
- The underfunded K-12 school system will get a revenue boost from the wind energy firm, and the local children will have a brighter future.
- Together we (community and our wind farm) are going protect the environment and reduce dependence on foreign energy, contributing to national security.

The most significant economic impact occurs after community cultivation, during the construction phase. These expenditures provide the biggest jolt to a local community's economic base. The turbines, the community organizing, procurement of land, synchronization of incentives and investment, link to the grid, and labor costs add up to tens if not hundreds of millions of dollars.

Millions of dollars are infused primarily into rural taxing bodies, which proves to be particularly useful to disadvantaged, underfunded rural school systems. Parents are content knowing their children are receiving a competitive education, and school system employees are gratified knowing they have enhanced job security: "We are focusing heavily on math and science. Our kids are going to get a top-notch education on par with what they get in Normal [Illinois]."

That said, much is hidden from the outward discussions of community benefit. The impacts of a wind farm are, by and large, up front and economic. Wind energy developers do not shy away from this fact, but they seem to conflate the short-term impacts with long-term economic gains. Plus, the wind energy developers are never wholly transparent; they convey only some of the economics of the wind firm, never producing total profit projections. The public is excluded in the governance of wind energy in a way that could advance long-term community development—in both the co-op and IOU models of ownership.

The long-term operations phase, while contributing to the local tax base and land-lease payments, provides relatively few jobs (though such jobs are typically well paid). Revenue generated from wind and land resources is typically absentee managed, resulting in capital flight concentrating wealth elsewhere. Additionally, longitudinal analyses are unavailable since the industry is in its relative infancy; this is problematic for communities pressured into making long-term development decisions (land-lease contracts for turbines can range from twenty to fifty years) based on short-term return-on-investment projections.

The wind energy developers are not in the local communities to create change per se. The developers are there to build buy-in to deploy an enterprise contributing back to core organizational values (profits for IOUs or enhanced services for cooperatives). It falls to the actors within the firm to make use of the generated resource of the wind farm. It is doubtful that significant social structural change could come from the building of wind farms. In fact, the wind farms probably did more to maintain the local status quo (though to varying degrees) than to challenge it.

Referring back to the topic of ideal communities for wind energy development, it may not be happenstance that two communities in Ward and McLean Counties have strong growth coalitions. A system can be structured to prevent individual and collective action for purposes of atomization or centralized power (or to subvert stifling social centers by entering into voluntary collective action arrangements elsewhere). Growth coalitions indeed utilize such practices to enable economic development initiatives.

A rural, agriculturally based growth coalition located in a wind-rich community is quite attractive for wind energy development interests. The economic aspects of wind energy development were desirable to the groups that play an influential role in growth coalitions. The wind energy firms in both of these communities link to cliques with growth coalition attributes. Property owners could lease their land to extract financial value from the wind, and the tax burden of property owners would be diminished as wind energy occupies an increasing share of local tax liability. Communities with agricultural commodity orientations allow a more straightforward pitch: the concept is not foreign to landowners who are used to harvesting their land for a living.

When the growth coalition helped lock in long-term leases for the wind farm, many community benefits are forfeited for the life of the wind farm (thirty to fifty years). Perhaps this is why many respondents were muted or underenthusiastic about operational impacts. The school official in McLean County, excited about the financial boost to the tax base, noted: "It's too bad . . . that we're probably prepping these kids to move out of here, though. There just ain't anything to keep them around." This statement highlights the lack of substantive change that empowers individuals to govern themselves within these communities. Sure, the wind farm provides resources that allow for the development of human capital through locally established government institutions, but inclusive self-governance and public entrepreneurship take a back seat to narrow financial and material outcomes it brings. The actors involved in the wind-energy development process seemed to represent a cross section of the local status quo regarding age, gender, race, and class; very little seemed to be done to diversify engagement.

The potential benefits of public energy governance are numerous. If the community is involved in the governance process, it could better utilize the wind farm for public pursuits. Outside of the obvious ramifications of distribution of resources, one must consider the physical characteristics of a wind farm. Properly sited, a wind farm could be used as a nongovernmental limit to the capacity of a community to sprawl, serving as an instrument of greenbelt and farmland preservation. Public engagement might result in higher community self-awareness. If private IOUs occupy the land, then the wind regime cannot be further harnessed with a new community-owned wind farm. Once a wind farm is built, saturating that wind regime, additional or alternative models become impossible. Communities must be careful to choose whether the ownership model introduced locally is the ideal community partner in the decades to come. Perhaps the community might collectively alter its calculus if the actors are aware that wind resources are indeed exhaustible.

Wind energy development certainly has an impact on the social structure of the host community. Wind energy development, at this current stage, is more amenable to the status quo economic development than to empowering community development. It becomes clear that communities need to be prepared for wind energy development up front to negotiate better terms. Once in construction and operational phases, the community benefit is predetermined. Individuals must be able to work collectively, build new institutions, and challenge centralized power by aggregating the existent social hierarchy or having the tools necessary to create their own social structure. But does the insertion of a given type of organizational model impact the type of community development outcome?

How Does the Ownership Model of the Wind Energy Firm Affect Community Development?

The question of ownership over the wind farm and ownership's resultant community development impacts requires peeling back of the layers of governance and the influence ownership has on governance mechanisms. Such an analysis helps assess not only how the firm is stewarded but also its resulting robustness and how the community is engaged in that process. Many actors within both types of firms perform governance, but the owners themselves rarely partake in direct governance or stewardship outside of the occasional shareholder or member-owner meetings. McGinnis and Brink (2012, p. 1) elaborate further on the concept of stewardship: "The term 'stewardship' . . . refer[s] to the practice of managing common resources in a way that ensures the continued availability of that resource to future users. The problem is a classic dilemma: each individual has an interest in extracting as many resources as possible and hoping that someone else will pay the costs of replenishment or maintenance. In the absence of effective stewardship, the commons will be destroyed."

The IOUs certainly make greater promises, conveying the benefits to individuals and the community in financial terms while avoiding direct discussions of potential pitfalls; the IOUs are not as concerned about their role in community stewardship and governance as the cooperative firm. A further point of consideration is that a firm whose substantive controlling interest comes from venture capital investors (IOUs) poses a risk for long-term sustainable operations. What if the market collapses or the firm goes bankrupt? What are the reasonable chances the IOU will make an effort to transition the relationship between the community and the wind farm in a mutually beneficial manner? Is the IOU the optimal steward for a community's wind regime?

What is needed is a tool to assess institutional features and how those features interact with the institution's host community. The Ostrom design principles (ODPs; see Table 1.2) serves as a useful diagnostic in assessing institutional robustness.[5] The categorization of institutional features provides the analyst with the tools necessary to identify shortcomings in institutional design and the institutional fit with the resource it stewards.

This research applies the diagnostic to discover whether and which design principles are being used by the firm and if they are helping to support the wind regime and community field as a robust resource, focusing on two individual organizations from within the broader system. The diagnostic allows for an analysis of the robustness of the firm, which helps illuminate how the firm interacts with the local community, the institutional logic guiding the firm in this process, and the governance features integral for community development.

For analytical clarity, *robustness* is defined as "the ability of human-constructed systems to remain functioning even after experiencing an exogenous shock" (McGinnis, 2011a, p. 177. Importantly, this allows us to link institutional robustness features to economic and social externalities. The question of how the ownership model of the wind energy firm affects community development becomes more relevant. With the ODPs in hand, a robust institutional analysis better links institutional design and practice to community governance and development processes.

Assessing the Design and Operational Impacts of the Wind Farm on Community Economic Development

This chapter has already discussed the broader IOU and cooperative systems and their regulatory environments, as well as the generalizable elements of the wind farm and community interaction. How do the two case studies inform our understanding of the role of ownership over operational wind energy firms and their interaction with the host community?

Evidence from fieldwork in the two case studies demonstrates that a cooperative wind farm offers some advantages not likely to be found with an investor-owned wind farm. These advantages stem largely from participatory governance linked to ownership rights and the nested systemic governance designs of the cooperative firm, which do not exist in the IOU model (IOUs are simply not structured in such a manner). What follows is a defense of these findings through an institutional analysis of the two wind farms, via the ODPs.

> ODP 1, *Clearly defined boundaries*: "Individuals or households who have rights to withdraw resource united from the . . . resource must be clearly defined; . . . the boundaries of the . . . resource must be well defined."
>
> ODP 7, *Minimal recognition of rights to organize (local autonomy)*: "The rights of appropriators to devise their own institutions are not challenged by external government authorities."

The boundary of interest here is the community's wind and governance regime. Importantly, how does an operational wind farm's ownership model impact a community's capability to alter the operational features of a wind farm, and how does that wind farm reciprocate?

A wind farm is, at its simplest, a collection of turbines serving as extractive devices. The infrastructure is sited across vast tracts of land (now limited in its productive capabilities), accessing and extracting consistent wind resources and converting it to electricity, which it transmits to the grid in exchange for money. The monetary value of wind energy depends on many factors (reliability and intermittency of wind, market demand, subsidy, cost relative to other fuels, etc.), yet the financial returns were not a part of the conversation in the community—community members were simply left out of the discussion of the allocation of surplus revenues.

It makes sense for the IOU to divert attention from a dialogue of cash flow. If the IOU can control the information available to the community while generating trust and buy-in among key decision makers, it is in a superior position to lay out its terms for development within a host community. The IOU works diligently to disengage potential oppositional issues, trumping up the financial benefits to landowners and the infusion of resources to symbolically important taxing bodies. Correctly performed, swiftly acquiring community acquiescence enables the IOU to optimize profits and revenues to be estimated over the lifetime of the wind farm. Once secured by contract and force of law, the fully operational wind farm becomes an attractive hedge for large financial interests.

Community actors must be cognizant of the fact that the IOU seeks to limit broad community participation. Education provided by the IOU is not an act of empowerment as much as another form of marketing or advocacy seeking acquiescence. It desires to do the minimum necessary to see a wind farm through to operational status. Speculative contracts with landowners are sought far in advance of a commitment to build—albeit with a promissory fiscal return should the wind farm be built. The processes as utilized by the IOU do not engage community members in the operational stewardship of the wind farm. Once landowners have approved a contractual agreement on IOU-ordained terms and local government has approved building and zoning permits, the community has little leverage to enjoin in governance (they have not been conferred rights since they do not have the privileges of ownership). This unfortunate practice can result in unforeseen criticism and harm the public sentiment so badly needed to support public policy leading to the conversion of renewable energy.

Take, for example, the issue of grid connectivity. According to a wind energy developer, it is not uncommon for some actors to demand locally generated wind energy be distributed and consumed locally as well (Nov. 9, 2011). Some residents in McLean County were upset that the electricity generated from Twin Groves would, in essence, be exported for use elsewhere. The perception was that a wind farm is perfectly capable of being used for local energy needs, so why should Horizon sell the electricity elsewhere? This general perception demonstrates a fundamental misunderstanding of how energy generation on the grid functions and the problems that arise when we exclude stakeholders from stewardship: confusion or ignorance can lead to allegations of impropriety, diminished trust, and opposition to the project.

One can envision scenarios where other, similar conundrums arise. With enough consistent, repetitive retelling of such friction going unchecked, these stories could engender a myth of wind energy development as community exploitation and misappropriation. Rapid, mass conversion from "dirty," exhaustible sources of energy to clean renewables could easily be impinged in such a scenario.

Cooperatives do better at engagement, distribution of resources, and long-term participatory engagement in governance. Basin and VEC avoid overcome the engagement problem faced by IOUs through intentional design and practice. The nested-systems approach of electric cooperatives means that any electricity produced by Basin for distribution by Basin's member cooperatives is for the member cooperatives' consumer-ownership; the boundaries are more logical (the wind energy turbines connected to the local transmission lines technically do feed back into the local distribution network). The control of these systems and the engagement of members are where ownership contributes to the furtherance of knowledge.

The electric cooperatives have many incentives for educating the community to better understand electric and wind energy governance and stewardship of the electrical system. Cooperative institutional logic is informed by the ICA cooperative principles and values (CP; see Table I.1) (http://ica.coop/en/what-co-op/co-operative-identity-values-principles). Of relevance here is CP 4, autonomy and independence: "Cooperatives are autonomous, self-help organizations controlled by their members. If they enter into agreements with other organizations, including governments, or raise capital from external sources, they do so on terms that ensure democratic control by their members and maintain their cooperative autonomy."

The cooperatives are rooted within the community (though Basin, as a super-G&T, to a much lesser degree). The board is composed of those who live within the service territory and consume the goods and services (and the staff likely is as well). Whatever the cooperative does in the community will affect those directly governing the cooperative, so they are incentivized to operate in the community's best interest. Furthermore, it is highly likely that the cooperative would not want to be seen as a destructive force.

The IOUs are part and parcel of a system that is enabled by the federal government to grow the share of wind energy generation in the United States. As discussed in earlier in this chapter, community ownership of wind is a virtual impossibility without active pushback by community development interests. Electric cooperatives serve a critical role in pushing their communities to actively engage in energy governance; without these cooperatives, public energy governance would be virtually nonexistent. Electric cooperatives are actively involved in defining the boundaries of their responsibilities toward these ends.

The cooperatives indeed pulled from the same proverbial playbook as the IOUs regarding appeals to individualism and communitarianism, but there was a stronger element of social responsibility in how the cooperatives engaged the community by appealing to verifiable, tangible outcomes.

First, the electric cooperatives provided in-depth educational presentations and materials. Basin not only was extremely sensitive to the known oppositional issues but also was very careful to acknowledge potential future dilemmas, such as the saturation of a localized wind regime that could limit capacity for wind energy development. Yes, this did help to engender more profound trust, but in presenting on such topics Basin risked its interests by identifying potential pitfalls.

Second, the electric cooperatives actively acknowledged that landowners should seek out their own third-party legal representation. Basin offered all prospective landowners legal fee stipends to retain legal representation to review the fairness of its contracts. Basin also suggested the landowners pool the stipends together to enhance the legal representation. Again, Basin took a risk by providing the landowners a venue by which they could work collectively outside of Basin's control and potentially mobilize to demand better contractual arrangements.

Basin and VEC have well-established mechanisms to allow for longitudinal governance over the wind farm. These mechanisms have been transmitted from among their cohorts in the electric cooperative sectors (discussed further below).

ODP 3, Wide participation in collective choice: "Most individuals affected by the operation rules can participate in modifying the operational rules."

The IOUs' narrowly designated boundaries, established to maintain centralized command and control, translate into a scenario where participatory collective choice mechanisms are mostly cut off after the development phase. Horizon developers made it very clear that the executive team would take a hands-off approach, expecting the development staff to produce optimal returns. Horizon developers sought the path of least resistance and desired to use monetary incentives to seek acquiescence from the community. Their time frame was short: do just enough to get the turbines in the ground, and the job was essentially done. The development staff would then pack up and head to the next field site; absentee ownership could then kick in. In this manner, IOU wind farms are duplicating many of the features of absentee coal-fire power plants that had historically atomized their host communities (Adamson, 2008; Freudenburg, 1984, 2008).

Governance was not so restrictive with the cooperative, which are not motivated by quarterly performance-reporting metrics. Cooperatives are oriented toward member needs. Cooperatives leaders see themselves as existing in perpetuity, benefiting current and future generations. The board and operational teams live and work within or near the service territory. What they do unto their service territory, they do unto themselves and their friends, family, and peers. The actions of the cooperative interact with the community, and the community also interacts with the cooperative. Electric cooperatives then exhibit a much longer time horizon and planning perspective than their IOU counterparts.

CP 1, Voluntary and open membership: Cooperatives are voluntary organizations, open to all persons able to use their services and willing to accept the responsibilities of membership, without gender, social, racial, political or religious discrimination.

CP 7, *Concern for community*: Cooperatives work for the sustainable development of their communities through policies approved by their members.

These two CPs inform how the cooperative engages the broader community, pushing the cooperative to seek venues for broad-based participation, as well as to ensure there is a net community benefit from its operations. Take, for, example Basin's and VEC's actions. Basin is quite cognizant of the potential for positive economic externalities reaped by the community hosting its wind farm. It makes sense that it could situate any significant investment within its service territory and within proximity of a member cooperative: VEC. VEC led the community effort to develop the wind farm, which fostered positive social capital from the community toward VEC. VEC's approach also enhanced Basin's standing among actors in Ward County.

Merely citing the infrastructure within a cooperative's service territory is not enough to claim broad participation by the member-ownership in governance over the wind farm. The electric cooperatives in this study are not complacent organizations. The board and executive team offer many incentives to promote active civic engagement by the staff. During the floods, staff members were offered paid leave to help mitigate the damage. Furthermore, VEC invests plenty of organizational resources in driving up turnout at its annual general membership meeting, which has become a community fixture. Although these examples are somewhat one-way participatory mechanisms, they do transmit valuable information back to the organization, which the leaders can then absorb and analyze to assess the need to recalibrate features of the cooperative. But these mechanisms do not necessarily empower community actors to govern.

VEC is also engaged in fostering new institutional leadership. The member advisory committee has provided an outlet for interested cooperative member-owners to actively engage in governance over the electric cooperative and participate in decision making. This practice carries a risk of disruptive member-owners taking advantage of an open door to the organization. It also carries the advantage of information sharing, trust building, and the generation of future leaders for the cooperative.

ODP 4, *Monitoring*: "[A] Monitors are present and actively audit . . . resource conditions and appropriate behavior; [B] monitors are accountable to or are the appropriators."

The electric industry as a whole has a relatively robust monitoring regime for the ongoing operation of the electric grid. The ISOs, FERC, and NERC serve the role of monitoring the grid for early warning signs, as well as vetting new connectivity to assure ongoing operations.

The operational IOU wind farm does not directly report much if any information to its local host community. IOUs provide reports primarily to regulatory agencies and to the investor-ownership. Performance and health and safety information can be found primarily online through the websites of governmental regulatory bodies. Unless otherwise specified in the terms of local community government or land-owner lease contracts, the IOU wind farm need not report additional data above and beyond what is required by law.

Electric cooperatives mostly perform the same reporting as IOUs and report to their member-owners and cooperative associations. The electric cooperatives in this research see this feature as linked to CP 5:

> CP 5, education, training, and information: "Cooperatives provide education and training for their members, elected representatives, managers, and employees so they can contribute effectively to the development of their cooperatives. They inform the general public—particularly young people and opinion leaders—about the nature and benefits of cooperation."

Oddly, electric cooperatives take it as given that they are unregulated; they seem to be unregulated by the government only regarding their business-to-business features as it relates to financial allocation and shared resources. The reality is that they are extremely regulated at a number of levels. Electric cooperatives report to their member-owners, at a minimum, once per year at the annual general membership meeting. In the case of VEC, its member advisory committees provide another venue for in-depth, intimate discussions of the cooperative's robustness. The cooperative must also report to their cooperative support system lenders on their fiscal performance, as well as to their marketing cooperative, Touchstone, on the integrity and quality of the member-owner services.

Electric cooperatives are highly regulated and monitored entities at many levels, by virtually all of their stakeholders. Not so with IOU wind firms. The lack of diversified stakeholder monitoring in the IOU firm calls into question the robustness of the IOU's public service orientation relative to cooperatives.

> ODP 5, Graduated sanctions: "Appropriators who violate operational rules are likely to be assessed graduated sanctions (depending on the seriousness and context of the offense) by other appropriators, officials account to these appropriators, or both."

ODP 6, Dispute resolution mechanisms: "Appropriators and their officials have rapid access to low-cost local arenas to resolve conflicts among appropriators or between appropriators and officials."

Both ownership models will have spelled out penalties between landowners and the firm, as well as local government agencies and the firm. The IOU, however, differs in that the sanctions community members can leverage against it will be limited (the IOU is absentee). Plus, the IOU is more likely than the cooperative to insist on contractually binding third-party dispute resolution mechanisms, such as arbitration, and the IOU has more resources available for sidestepping local governance and putting disputes in more favorable regional or national actors, such as the court system or energy regulatory agencies.

Electric cooperatives could undoubtedly operate in an absentee manner (though it would seem to be against their institutional logic) and could have their member-owners sign a waiver for court access and rights for disputes processes. However, in the case of Basin, that did not occur; the member-owners certainly could take given disputes to court should the need arise.

Cooperatives have additional sanctioning and dispute resolution mechanisms. Simply enough, member-owners of a community could exert collective social pressure on actors within the cooperative to resolve disputes. Member-owners can also attend VEC or Basin general membership and advisory meetings, where their voices are heard again. And, should it come to that, the member-owners can call for a special emergency meeting, the board can fire staff, and the member-owners can expel members of the board. The community certainly has more options for dealing with a problematic electric cooperative than with a similarly problematic IOU.

ODP 8, Nested enterprises: "Appropriation, provision, monitoring, enforcement, conflict resolution, and governance activities are organized in multiple layers of nested enterprise."

All wind energy firms are embedded within a nested system. The electric grid is, at its base, a networked system whereby actors must minimally be linked to government regulation and oversight, as well as maintaining connectivity to financiers on matters related to investment partnerships and future investing opportunities. Moreover, if the accounts of developers at Horizon were accurate, wind energy developers stay connected with one another to share best practices for community organizing, as well as infrastructural build-out.

Electric cooperatives nest within the same network of actors but do so by way of their own polycentric, federated system. Cooperatives are by design a nested enterprise, and that nesting is utilized to enhance the cooperative's local, democratic value orientation. The institutional logic is again informed by the co-operative principles:

> CP 6, Cooperation among cooperatives: "Cooperatives serve their members most effectively and strengthen the cooperative movement by working together through local, national, regional and international structures."

Electric cooperatives are by law and institutional design governed by their member-owners. They are also nested within their community. Unlike the IOU, there is a recognition of participatory rights to governance and monitoring. Taken together, this means that electric cooperatives operate as a gateway, an access point to active participation in the governance of wind energy and the electric grid. Taken on the whole, the electric cooperative provides a venue by which citizens may actually participate in democracy, aggregate resources to enhance the local cooperative's service offerings, and foster a system that mitigates the public service paradox through consumer engagement in provision and procurement processes.

Officials from the two cooperatives and the cooperative actors in the associations talked in terms of evolving toward a complete systems approach. The electric cooperatives in this study exhibited an aggressive desire for vertical integration ("from mine mouth to meter") so that they might enhance their self-governing capacities through control over market share. Vertical integration of the supply chain by the G&Ts is coupled with the federated structure in an attempt to control market volatility and organizational resilience as much as possible to meet this end.

But it is not just a market orientation for the electric cooperatives. Through association with the national apex organization, NRECA, the cooperatives have created a wide variety of support systems for the electric cooperative sector. The electric cooperative system is incrementally moving toward being able to self-finance, has created a marketing brand (Touchstone), and is aggregating individual, organizational needs into group purchasing negotiations to drive down overhead costs on services like health insurance and pensions. The cooperatives are in a persistent state of network maintenance.

The IOUs prefer targeted, isolated specialization, as well as the externalization of management of the grid infrastructure. IOUs may be nested within the electric energy system, but they are not necessarily nested within their host community. By design, the IOU actors act as nodes on a network, closed to community engagement.

In contrast, even a corrupt cooperative offers latent capacity not built into the IOU firm. The value the cooperative may offer to the member-owner is undoubtedly enhanced by the nested nature of electric cooperatives; the cooperative firm has much more in common with its host community than the IOU does. It is left to the actors within the local-level electric cooperative to transmit the value into something meaningful for the local community. The communication of value is where adherence and commitment to democratic governance become challenging.

ODP 2, *Congruence with local conditions and fairness*: "Congruence between appropriation and provision rules and local conditions: [A] Appropriation rules restricting time, place, technology, and/or quantity of resource units are related to local conditions; [B] The benefits obtained by users from a . . . resource, as determined by appropriation rules, are proportional to the amount of inputs required in the form of labor, material, or money, as determined by provision rules."

Analysis of the firms failed to find evidence wherein the cooperatives or Horizon dramatically altered the social structure in a way that could be identified as community development. Horizon sought to harness those with social and political capital to influence development. IOU wind energy developers sought approval from local government regulatory agencies and contractual agreements with landowners. The desire was to limit overall tax burden, limit community enticements, and maximize the subsidies to guarantee the largest marginal payoff possible. Up front, the community and wind developer may be in alignment with their values orientation, but as time goes on, the community has few if any options to renegotiate the terms of the contract. Once a wind farm is constructed and operational, the community is stuck with that infrastructure for thirty to fifty years, and Horizon's maintenance of its local community network basically shuts down.

The electric cooperatives come from the local community and seek constant alignment with the broader value orientation of the community through processes seeking consensus. The electric cooperatives are helping both to contribute to the maintenance of the existing social structure and to build individual and collective capacities. Board members, member-owners, and operational staff are interacting through standard channels created and sustained by the cooperatives and the broader community. The product of these interactions seems somewhat static from a structural change perspective; much like the IOU, the cooperatives seem to be engaged with their local growth coalitions.

However, recent developments within NRECA, the U.S. cooperative movement, and the ICA could stimulate these actors to participate more explicitly in local cooperative development. NRECA created the Twenty-First Century Committee in reaction to the ICA's Year of Cooperatives celebration to assess how cooperatives could get back to their core mission of democratic member participation. According to actors involved with the committee's work, there is a push for system-wide realignment with the CPs (co-op association official, Sept. 8, 2010). For this analysis, the relevant cooperative principles are as follows (ODP 3 addresses a sizable portion of CPs 1 and 7):

CP 1, *Voluntary and open membership*: Cooperatives are voluntary organisations, open to all persons able to use their services and willing to accept the responsibilities of membership, without gender, social, racial, political or religious discrimination.

CP 2, *Democratic member control*: "Cooperatives are democratic organisations controlled by their members, who actively participate in setting their policies and making decisions. Men and women serving as elected representatives are accountable to the membership. In primary cooperatives, members have equal voting rights (one member, one vote) and cooperatives at other levels are also organised in a democratic manner."

CP 3, *Member economic participation*: "Members contribute equitably to, and democratically control, the capital of their cooperative. At least part of that capital is usually the common property of the cooperative. Members usually receive limited compensation, if any, on capital subscribed as a condition of membership. Members allocate surpluses for any or all of the following purposes: developing their cooperative, possibly by setting up reserves, part of which at least would be indivisible; benefiting members in proportion to their transactions with the cooperative; and supporting other activities approved by the membership."

CP 7, *Concern for community*: "Cooperatives work for the sustainable development of their communities through policies approved by their members."

Realignment with the overall bundle of CPs could lead to significant positive outcomes. North Dakota has a rich history of cooperatives participating in economic and community development. The so-called co-op fever of 1991 had substantial participation by North Dakota electric cooperatives, including VEC and Basin (Patrie, 1998). This story is again latent capacity for community economic development emanating from cooperatives; all it takes is the right conditions and catalytic actors to drive the cooperative toward community development engagement.

Currently, VEC is both helping contribute to the maintenance of the existing social structure (which is critical in an environment of a rapid influx of in-migrants, which could unintentionally lead to a great deal of chaos) and building individual and collective capabilities for engagement in community governance. The current leadership of the electric cooperative is looking at nurturing the next generation of talent to take the helm through the member advisory committee.

Steadfast adherence to the preferences of local actors does not come without some pitfalls. Cooperatives exhibit enormous potential to influence the larger system. Officials at the DOE made it clear that NRECA and the rural electric co-ops have immense political clout, which is being used for initiatives deemed worthy by key leaders in the sector. Electric cooperatives have not been actively pushing awareness of the need to convert from coal to renewables such as wind, which means member-owners are not mobilizing en masse for rapid conversion. Electric cooperative clout is not used to grow the cooperative movement or community development capacity as much as to maintain legacy infrastructure and to limit change, especially the shift from coal to renewables. Basin, for example, is only looking to build its share of renewables within its portfolio, not to convert it entirely. Basin projects it will not need new electric generation capacity until 2018 (cooperative official, Aug. 18, 2011), meaning Basin, with the greatest potential of all energy producers to develop North Dakota's wind regime, will remain relatively idle in this area, reducing cooperative participation in renewable proliferation, leaving the gate wide open for IOUs to saturate the wind regime.

Rural American communities need their electric co-ops more than ever. But right now, electric co-ops are suspiciously silent.

CHAPTER FIVE

Why Not Policy from Below?

Rural communities can govern sophisticated enterprises and resource systems to enhance collective well-being. As demonstrated in the evidence presented in this book, rural communities equipped with the proper set of institutions and leadership can serve as platforms to steward wind energy resources toward a liberating community economic development. This is encouraging.

What does a comparative case study tell us about the potential for rural communities to thrive in an era of economic decline and democratic deficiency? Are community-based organizations situated at the forefront of serving these rural communities the appropriate vehicles for enhancing economic democracy? Moreover, does economic democracy have an appreciable advantage over unfettered market economics?

I conclude with some final analyses and perspective, based on the four major themes that run throughout this book: top-down economic development, community participation, an inclusive approach to wind energy development, and grassroots-oriented policy. I close this study with some thoughts on how rural Americans can engage with wind energy development to mobilize a liberating approach to community economic development, with community leaders playing a central role, and the challenges that these leaders should consider.

Wind Energy Development: Top-Down Public Policy and Economic Development

U.S. renewable energy policy is premised on two broadly desired outcomes. First, government mandates, subsidies, and other incentive mechanisms are crucial toward the furtherance of and conversion to clean renewable energy. Second, the rapid build-out of this next-generation energy infrastructure will confer many community economic development benefits to the host communities. I found that U.S. wind energy policy falls short of these policy objectives.

Government incentive structures serve the purpose of (a) creating market demand for wind energy, (b) reducing the wholesale purchase price, (c) driving down consumer cost, (d) decreasing investor tax burden, (e) increasing the return on investment in order to entice investors, (f) creating a stable hedge against market volatility, and inevitably (g) increasing the rate of growth in the wind energy industry. The policy then "bakes in" top-down market orientation, relying on profit-driven entities for wind energy proliferation. This invisible-hand orientation places a level of faith in profit-driven actors to advance policy, which is advantageous for IOUs but does little for not-for-profit enterprises like cooperatives. Put another way, U.S. energy policy privileges a handful of individuals in the investor class, failing to acknowledge the 42 million rural Americans who own the electric co-op sector and who could serve as a small-dollar investor class for wind energy.

The electric sector, like an increasing number of public goods and common-resource regimes, operates under a homogenizing market orientation. Many community-based organizations have adopted a values orientation in line with the interests of the investor-owned firm. This is in part because the playing field has been tilted in favor of profit-oriented firms.[1] The provision of public goods is viewed by policy makers as optimally distributed via market mechanisms, resulting in policy emphasizing competition, individualism, and rent-seeking. The replacement for "the state" with "the market" presupposes a false binary, thereby limiting choice, cultivating a system that is guided by market-oriented perspectives that shape the governance structure, leaving a community's interests in the governance structure unaddressed as a matter of policy.

The prescribed market-based solution sidesteps debates of the rent-seeking tendencies of the IOU and the relevant behavioral propensities toward its host community (crowding out entrepreneurship and operating in an inefficient, command-control manner). When public discourse is inhibited, government policy makers, lacking critical self-reflection, conflate means (markets and competition) as ends (choice, self-governance, and community well-being). The wind energy system is in a self-perpetuating loop, getting stronger and stronger despite indicators of impending socioeconomic disruption (economic bubbles, energy security, and global climate change).

Community-based enterprise is unlikely to thrive under such structural barriers. Entrepreneurship is limited. Dependency on hierarchy and centralized authority is reinforced. Vulnerability becomes a feature, not a bug.

The "if you build it, they will come" wishfulness of the second premise, that wind energy development will naturally result in community economic development, wrests on unverified trust and assurances. The policy lacks intentionality by not requiring minimal protections and benefits to the local community (e.g., the law is often silent as to who is responsible for retiring decommissioned wind energy infrastructure). Missing is language mandating educational programming for landowners hosting the turbines and the local officials making important zoning and taxation decisions. Missing is information about the profitability of wind energy.

Missing are specified opportunities for the community to participate in the profitability of the wind farm. The information asymmetries are incredibly lopsided and utterly disempowering; the wind energy firm holds all of the cards, and the community must be aware of the right questions to ask. Advocacy for wind energy is, at best, a domesticating form of education and, at worst, base exploitation.

Withholding broad-based community benefits is often justified by the fact that most wind energy firms are de facto private in their investor-ownership, and free enterprise should not be required to release proprietary trade secrets. But the consumption of public resources (subsidies and tax credits, and readily available public infrastructure), along with the distributed spatial features of the wind farm resulting in sprawl across a community, means the burden should be on the wind energy firm to fully inform the community of the challenges and opportunities in hosting wind energy infrastructure. As demonstrated by the case of Horizon in McLean County, Illinois, the wind energy firm was willing and able to utilize surplus revenues to secure community acquiescence through a common fund, but the fund was never offered because the community did not know to ask for it. Communities would be wise to organize and protect their best interests as opposed to waiting for government regulators to provide for their well-being.

It is not a stretch to see how community members could perceive the actions of the wind energy firm as manipulation of public goodwill. The current policy regime privileges the IOU wind energy firm over the community, meaning wealth extraction is privileged over community economic development. Public policy's lack of specificity on the issue of community economic development may eventually stunt the growth in wind energy. In fact, one wonders if the extractive nature of wind energy actors may be what is catalyzing wind energy development opposition, harming policy adoption. In this manner, the policy is suboptimal for achieving the two premised outcomes.

Community Participation: Passivity or Intentionality?

It is clear that the wind energy development phenomenon should not be presumed to be a net benefit—nor a detriment—to a host community. However, reliance on spontaneous development outcomes is a naive approach that presumes the mere introduction of a major development project would result in positive community economic development. The positive externalities of a wind farm depend on many factors (how it is operated, the underlying logic of the ownership model, who governs the firm and their underlying motivations, and within what framework of rules under which the firm functions). The community benefit is a result that can never be guaranteed, but it can be better controlled with some level of regulation or community intervention. One cannot overemphasize the importance of intentionality in the planning process. Communities must mobilize and account for their interactions with wind energy firms.

I observed the downside of passive development, of allowing "markets" to "happen" as opposed to intentionally designing or controlling for development outcomes (taming the invisible hand of the market). While proponents of free markets may be aghast at those comments, one cannot deny that private enterprise does its best to control for exigencies; why should we deny these facts? So why should communities maintain a passive posture in the face of such projects? Wind energy development receives government assistance with virtually zero strings attached, as if the consumption of public goods confers little if any responsibility to provide for the commons (i.e., community, grid, and national energy policy). Wind energy developers are embedded in the same networks as finance, coal, oil, and natural gas interests, and many of their representatives are part and parcel of those industries. The investor-owned firm is not legally or normatively expected to account for the public interest and is indeed presumed and allowed to game the system where possible. As an example, there are increasing incidents of investor-owned wind farms refusing to power down their generators when requested by the ISOs, due to the need to produce as much revenue as possible. This self-seeking behavior runs the risk of overloading significant sections of the grid, causing immediate crises for health care providers and emergency personnel; the unfettered private profit motive is indeed a threat to the robustness of the energy commons. To place absolute trust in the public-oriented virtues of the investor-owned firm is naive.

What was observed in this passive model of community economic development? First, the community-wide benefits are mostly front-loaded. The build-out of the wind farm requires a large number of laborers (hundreds to over a thousand) to erect the turbines and connect the wind farm to the grid. Most of the laborers come from outside of the community and are well paid, spending a substantial amount of their earnings on food, lodging, and other services in the community. This economic injection occurs during about six to eight months. Wind energy developers spend an inordinate amount of their boosterism emphasizing these short-term benefits as if they are permanent features of the wind farm. They are not.

Second, the long-term benefits are inert, doing little for the marginalized in a community. Locally rooted field staff number in the teens for a standard eighty-turbine wind farm, though those numbers are decreasing as monitoring technologies allow wind energy companies to outsource monitoring to a centralized command center, effectively cutting staff to zero. Local taxing bodies indeed reap the rewards of increased revenues (though the state austerity movements counterbalance these enhanced local tax streams). Property owners benefit from sharing the tax burden with new taxpayers and benefit further with long-term leases providing annual per turbine revenues of \$4,000 to \$9,000 on a quarter acre of land. These same landowners benefit from higher property values.

What is missing in the long term is a more general collective benefit.

- What is the effect of wind energy development on low-income residents?
- In those instances where renters must share their space with a turbine, do the benefits of hosting wind energy turbines accrue only to large landowners? Or are the property tax savings passed on from landlords to the tenants?
- How could the community capture the value created by these wind farms to break dependency and increase local self-governing capabilities?

The answers to these questions are of enormous consequence for hollowed out rural communities facing an end to hourly wage labor. Productive assets like wind energy could and should be discussed as a source of revenue for these communities.

Waiting on public policy alone is a risky hedge. Communities must consider organizing for intentional community economic development outcomes. This is not a new idea, either. According to USDA officials, there have been instances of ranchers working collectively to negotiate with wind energy developers for better leasing terms (Brockhouse, 2008). What if a coalition of landowners, civil society groups, and local officials banded together and connected with other communities who have worked with wind energy firms to mitigate the community information deficit and negotiate from a position of relative power? Communities could demand more favorable leasing agreements, call for a collective community fund, or create new revenue-generating opportunities for critical public services (e.g., broadband Internet, or a business start-up fund). A well-organized coalition could go so far as to predetermine how to situate the wind farm to use nongovernmental mechanisms to halt sprawl, creating a de facto green belt around the community, without intervention by a government planning agency.

Inclusive Wind Energy Development: Considering Ownership of Public Goods and Common Resource Regimes

This analysis provides much for key actors (developers, planners, and policy makers) to consider, particularly with considerations of the efficacy or usefulness of wind energy ownership in addressing the pitfalls of wind energy development. The questions of wind farm ownership become of central interest in that ownership is presumed to be a partial determinant of how actors navigate complex markets and policies at local, regional, state, and national levels and how those actors control for resource systems and better align values of the firm to the unique features of their community. Direct community ownership would hypothetically remove the barrier to citizen governance, enhance informational and resource (profits) flows, and supplant a potentially extractive enterprise with a potentially regenerative enterprise.

Energy policy involves how and why actors use essential resources to supply the grid, as well as how the grid infrastructure is shared and stewarded and influences the communities they touch. A central tension sits at its heart: on the one hand, following from the concept of the commons, the electric system is stewarded to varying degrees by every citizen, and every citizen has some form of governance responsibility over that system (paying the energy bill, conserving energy, serving on an electric utility's board); on the other hand, the American economic system has granted significant rights to private interests to extract natural resources, process them into usable "things" (goods or services), and profit from that process—often enhanced through government subsidy—as long as those private interests follow the rules established by the political process. Policy diminishes an opportunity for citizen governance (democratic administration) when we do not allow citizen-governed firms in the mix. The lack of citizen governance is partially a result of significant shifts in the orientations of government policy makers over the last couple of decades (observed in Reaganomics and Clinton-Gore's "reinventing government"). Government policy then confers privilege on the closed-door, for-profit, investor-owned firm with the presumption that the investor-owned firm will always offer the optimal policy outcome.

Robust public governance is about more than mere nonstate, market-oriented institutional arrangements. Aligica and Tarko (2012, p. 246), commenting on Bloomington School governance studies, remind us that "Ostroms' exploration led to the conclusion that the discussion on polycentricity is not just a discussion about multiple decision-making centers and monopolies of power, but also a discussion about rules, constitutions, fundamental political values, and cultural adaptability in maintaining them."

And yet the inherent conflicts between joint ownership, values, public participation, adaptability, and the exploitation of natural resources goes mostly unaddressed in U.S. energy policy. Incomplete, investor-friendly government policy harms alternative ownership models, creating a power vacuum tailored to and readily exploited by dominant investor-owned firms.

The way that a wind farm impacts its host community is found in the long-term stewardship of the wind regime and the resources generated by the firm; local ownership is one approach that seems to enhance community benefits (Kildegaard & Myers-Kuykindall, 2006; Tolbert et al., 2002). In this concern, a cooperative departs from an IOU on two significant points that should come to the attention of community developers and those advocating inclusive community governance. The first is that an investor-owned firm has many legal and contractual barriers isolating broad-based participation, whereas a cooperative serves as a point of entry for participating in governance over a good or resource. A community cedes many governance rights when it grants an investor-owned firm control over local resources. The investor-owned firm is typically absentee owned, meaningful interaction with the firm's central decision makers (executives, shareholders) may involve extensive coordination, or even outside arbiters (e.g., state or federal courts), to gain minimal interactions to discuss base concerns of the community. The points of entry are few, and the transaction costs of access are high. Community governance is further bounded, and community members lose voice.

Rural electric co-ops are frequently seen as community complements, benefiting their hosts. The rural electric cooperative sector has a history of empowering collective action, of people offering to pay for a neighbor's member equity, of farmers volunteering their time and equipment to dig post holes and string wire across vast terrains. Many of these early electric cooperatives went on to start telecom cooperatives, drinking and wastewater cooperatives, and credit unions. A few exploratory studies found that cooperatives, in general, outlive their corporate counterparts (Murray, 2011; Stringham & Lee, 2011) and root the community wealth through local ownership (Alperovitz, 2011). These are long-lived, durable organizations that engender public entrepreneurship and build community capital. If we are to have community economic development policy, it must account for cooperatives if the policy is to have any legitimacy.

This brings me to the second significant difference between IOUs and cooperatives: a member-owner has a right to direct participation in an electric cooperative. Even a corrupt cooperative must by law allow some minimal amount of participation by the member-owners. Any member-owner of a cooperative has a right to access governance mechanisms of the electric cooperative. By contrast, once an IOU wind farm is erected, governance is provided either through direct financial investment (requiring a sizable amount of money) or through the arcane regulatory apparatus of the public utility commissions, ISOs, FERCs, and NERCs. So while many electric co-ops must also answer to these regulatory bodies, electric co-op leaders are tasked to serve the local ownership. There are, of course, the performance reports and general membership meetings that are part and parcel of the electric cooperative model. But there are also direct democracy mechanisms, such as the ability to run for the board, participate in public advisory events, and personal interaction with the locally based staff and board members. These individuals are in the community—in the grocery store, in the city council meetings—meaning in many situations the co-op's leadership is visible and accessible to the public and can be held to account. In contrast, a fully operational IOU is not necessarily obligated to do so if it is not spelled out in the contractual details. That also means an electric cooperative can be governed or reformed from many angles. Reformation of an IOU requires a virtual social movement to change the laws, to alter its behavior, whereas consumers of an electric co-op have seemingly endless points of access and participation.

While an electric cooperative may cover a vast geographic territory (e.g., with a G&T), a member-owner located hundreds of miles away from Basin's headquarters can still access the central cooperative through the local distribution cooperative. Moreover, through this distribution cooperative's network, the individual member-owner can attempt to mobilize other distribution cooperatives under the Basin umbrella to influence change.

The study of VEC in Ward County, North Dakota, showed that the institutional logic of the cooperative is unlike that of the investor-owned firm. Electric cooperatives are by design (and legal statute) user-owner governed. They are also nested within the electric sector through a federated structure that provides residents access to a sophisticated, national-scale, sociotechnological system. There is a recognition of participatory rights to governance and monitoring. Having rights to participation means that electric cooperatives operate as a gateway, an access point to active participation in the governance of wind energy and the electric grid as a whole. In total, the electric cooperative provides a venue by which citizens may actually participate in democracy, aggregate resources to enhance the local cooperative, and foster a system that mitigates the public service paradox through consumer engagement in provision and procurement of goods and services demanded by member-owners. Even a stale electric cooperative provides additional value that an IOU would be reluctant to offer; namely, the very existence of an electric cooperative is at minimum latent community economic development and could eventually be used for empowerment.

VEC also demonstrates that an electric cooperative with engaged leadership can leverage the firm's resources for enhanced community economic development outcomes. VEC has kick-started many other local cooperatives and is very involved in the local civic activity and dedicated significant resources to disaster recovery efforts. In the case of VEC, it is clear that the leadership views the robustness of the co-op's longevity as linked to the health and longevity of the community. This view is an orientation not found as explicitly in the investor-owned firm and may even offer a competitive advantage to the co-op firm.

Referring back to underlying premises of U.S. renewable energy policy, if the proliferation of clean, renewable energy is the desired outcome, then it would appear that cooperatives must be factored into the equation. The participatory democratic governance and economic ownership features, coupled with the potential for community benefit, serve as significant trust building mechanisms. These mechanisms lend credibility in that the cooperative utility will better steward the surplus resources for community benefit while engendering the community trust necessary to build buy-in for the proliferation of wind energy capacity across the United States.

A Grassroots Public Policy? Communities as Governors of Their Own Destiny

Rural America is wind energy country, and electric co-ops are a potent vehicle for rural America to harness wind energy for community economic development. However, rural America's ally, the electric cooperative sector, seems to be in a deep slumber when it is needed most.

Again, public policy is mostly designed for creating market demand for wind energy, reducing the wholesale purchase price, driving down consumer cost, decreasing investor tax burden, increasing the return on investment in order to entice investors, creating a stable hedge against market volatility, and inevitably increasing the rate of growth in the wind energy industry. Co-ops are largely excluded and silent on the policy disparity. This exclusion is tragic in an era of corporate oligarchy, centralized economic dependency, and diminished democracy.

Historically, electric co-ops have been models of economic democracy. The electric energy commons is complicated and difficult for the average person to access; the IOUs are powerful, and government policy plays a commanding role in shaping the system. Cooperatives have demonstrated a long-term, successful track record for not only governing their share of the energy commons but also for maintaining their integrity despite the homogenizing tendencies of the system. In our recent history, as electric co-ops met their member needs they applied their entrepreneurial skills to build other democratic community-based organizations across numerous complex industrial sectors. While there is cause for concern, all is not lost, not by a long shot.

Forty-two million rural Americans own, govern, work at, and manage over nine hundred electric cooperatives, and the Cooperative Network, provides services at multiple, cost-effective scales in marketing, finance, employee benefits, and aggregated purchasing (see https://cooperativenetwork.coop/). The system has three sizable financial entities servicing their needs, CFC, CoBank, and the USDA's RUS, plus a total of over $40 billion in capital reserves and the collective crowd-sourcing capacity of the members within their service territory. Electric co-ops exist as a major engine of economic power. It is fair to say this is a sector that is punching below its weight.

I cannot overstate that institutional design alone does not wholly determine organizational behavior. All firms are governed and managed by often generous, sometimes self-serving, but always fallible human beings. Rural communities are rife with stories of the charitable entrepreneurs who "did good" for their communities. Moreover, there are numerous accounts of cooperatives engaged in behavior that some might call "uncooperative." However, rejecting co-ops due to the actions of one actor is akin to throwing the baby out with the bathwater.

The inherent power of the electric cooperative is that, by default, the cooperative incorporates users in the governance process. If the energy generator is cooperatively owned, the staff and the board steward the firm with the locality in mind. These individuals come from the service territory of the electric cooperative, providing a number of access points for member-owners to engage these direct stewards. Even then, these direct stewards can catalyze involvement, as is being done in the case of VEC's member advisory committee, which is fostering a new generation of cooperative thinkers, and entrepreneurs. The electric cooperatives certainly make an effort to educate their member-owners, as well as the general public (indeed, there is a renewed effort by NRECA to get its cooperative member base to educate the individual members and have them advocate for co-op–positive policy). The question is in the content and liberating intent of the education.

To address this, I briefly return to the specific case of VEC. While VEC is undoubtedly working to build its member-owner capability to perform community governance, cooperatives could also serve an influential role in co-opting the local community. One should be concerned that the efforts of VEC to build a civic culture may be a wash in the end due to its interactions with the growth coalition in the area.

The way a cooperative functions has important implications regarding broad-based wealth creation and community economic development. The governance mechanisms of the electric cooperative readily provide for an iterative approach to collective resource management whereby community actors may cycle through the cooperative and transmit new knowledge throughout the community. This matters for relatively minor operational issues, such as an understanding of utility bill. Additionally, major social dilemmas benefit from such engagement—global climate change and the intersection with energy policy in particular—in that these dilemmas will require a broadly based movement of engaged citizens at many levels, using a number of approaches to propose new, innovative solutions (Ostrom, 2009). An IOU simply does not see this as a core service to be provided to their community, or an obligation to work toward better energy and climate policy.

There is undoubtedly additional value provided by this cooperative-owned wind farm, and that additional value does seem to stem from the cooperative ownership model. The existence of a cooperative is not sufficient enough to claim community economic development. Embracing and amplifying the differentiating features of a cooperative (adherence toward core principles, enhancing member-owner engagement in participatory governance) moves the firm in that direction. In this way, it would seem appropriate to claim cooperatives, at a minimum, as having latent community economic development capacity.

The reluctance to claim cooperatives as institutionalized community development is appropriate. Claiming cooperatives as inherently "good" runs the risk of fostering complacency, limiting self-reflection, and degrading praxis. It also fails to hold the agents (board and management) accountable. Institutional analysts would be wise to avoid making a value claim absent of a number of parameters.

Action Steps

If one were so inclined to engage in economic democracy through one's electric coop, what might one do? First, cooperative member-owners must recognize and actualize their right to participate in not just their immediate co-op but also the cooperative network their cooperative is embedded within. This network is how a local cooperative provides any of its members a voice that gets heard, even in an isolated part of rural America, by providing an authentic gateway to a national network of co-ops that imparts economic and political power, at scale. Many co-ops across the sector fail to educate their members about their industry. Engage, ask, map out the assets and better understand how the local co-op operates in the open market.

The second step is to engage with management, with knowledge of the co-op sector in hand. Reaching out to the cooperative's management team is a simple yet powerful way to begin the process. Be open-minded, and probe before making any judgments about the co-op. One may discover a management team that has been doing remarkable community economic development work under the radar, waiting for the member-owners to seek more direct engagement.

However, there is a cautionary note: be wary of "domesticating" language and behavior that limits participation by the member-ownership and retains central control in the hands of management. I have seen such language arise in casual conversations with co-op management. When asked how one co-op compared to its investor-owned peers, the management gleefully boasts, "Well, of course we are high performing! We are a co-op after all!" But when asked about the co-op's contributions to its community and its role in empowering individuals beyond the baseline product offerings, a magical twist occurs: "Well, there's only so much we can do. We are a co-op after all!," implying that the manager is limited by some inherent deficiency in the cooperative ownership model.

The cooperative member-owner must be vigilant to the use of domesticating language that may portend a cooperative captured by management for self-seeking purposes. However, the co-op may also merely have had decades of malaise. The point is to operate in good faith and identify the precise problem as opposed to prescribing the wrong treatment for the wrong ailment.

Third, directly participate. Ideally, the management and board are open to working with you. You can participate, and member-owners have a great deal of power and authority in helping a cooperative that is in need of visionary leadership in community economic development. Perhaps you can participate in an advisory committee with the board or staff, or even seek out official positions as a board member in the next board election.

Once you have settled on a position of influence at an electric cooperative, you can then turn your attention toward the task of economic democracy. The co-op could look for economic opportunities to develop renewable energy for the general marketplace, thereby creating a revenue-generating program for the community. There are new renewable energy cooperative models, and nothing is keeping the electric co-op from investing in that sector, or kick-starting a new local enterprise (e.g., a solar or wind energy cooperative).

Fourth, participate in the Cooperative Network. The electric cooperative sector owns major marketing, finance, and pension ventures; these can be utilized to grow new cooperatives or provides services to those co-op sectors without a sophisticated and robust support network. Again, one must be constantly vigilant of domesticating discourse even amongst peers in the Cooperative Network. They may say their current strategy does not allow them to invest in something outside of their traditional operations, or even that the law keeps them from engaging in new endeavors. These individuals need to be reminded that they are tasked first and foremost with serving their ownership communities; instead of ceding economic market share to profit-seeking entities, our co-ops need to embrace a more aggressive tack to identify and meet member needs, instead of writing off new ideas and concerns. Also, if there are legal and regulatory barriers, the electric cooperatives have the backing of 42 million potential voters, a fierce lobbying force, and significant economic power to bring to bear. Instead of obstructionist, domesticating language, actors within the sector must enter into liberating dialogue that empowers.

There are also many convening arenas for the Cooperative Network that provide enormous economic and political capacity for electric co-ops, and other co-ops. These conferences and retreats are jam-packed with managers and board members. However, a community-oriented co-op member must observe and ask, where are the rank-and-file member-owners, and are their voices being heard at these conferences?

Fifth, on the topic of scale: "The nesting of cooperatives into a larger association of cooperatives allows for dynamic economies of scale, and the enhanced provisions of goods and services. This then strengthens the capacity of locally rooted electric cooperatives to concentrate their member orientation and Basin to focus its specialized service portfolio." Electric co-ops should harness scale to achieve bottom-up collective actions that, in the aggregate, amount to public policy. Advocacy coalitions could venture into territory controlled by unscrupulous electric co-op leaders to clear space for community voice and community benefit. Scale can also be leveraged to address inequalities wrought by the current investor-dominated system and further prosper economic democracy, particularly in those areas where investor-owned wind farms are seeking to exploit communities.

Scale could and should be used to guarantee that public dollars and policy are being used for the common benefit. We should take umbrage at the suggestion that electric co-ops are powerless; 42 million Americans own the sector and could be mobilized to apply pressure on their local, state, and national political figures. They could also become actively engaged as citizen investors, pooling their monies to invest, own, and generate revenue from wind and solar energy development.

Sixth, those who engage with the co-op sector should consider elevating the stories of empowering cooperatives as opposed to focusing on the few bad actors. Elevating stories of empowering co-ops and their leadership not only helps set the gold standard for ideal behavior among the cooperative business community but also helps set the standard for how investor-owned firms should operate. Why should processes of hegemony and isomorphism benefit the investor-owned perspective? Why can't cooperatives attempt to make investor-owned firms operate more like co-ops? By connecting these dots, we can push collective influence from the bottom up, to use co-ops as vehicles for social change, aggregate policy, and keep the destructive tendencies of rent-seeking organizations and individuals in check.[2]

Tocqueville's Democracy: Repositioning Community-Based Organizations as the Laboratories and Classrooms of Civil Society

A one-off study of two different ownership models of wind farms reveals there exists an opportunity for citizen governance (democratic administration), but that opportunity is diminished when we do not allow citizen-governed firms to participate in community governance and public governance at multiple scales. What is troubling is that, despite an abundance of cooperatives in the electric and other sectors, we do not readily observe initiative on the part of the co-op sector to take on this role.

Do electric cooperatives see themselves as vehicles for economic democracy, within a broader democratic movement? Alternatively, are they just utilities whose members really do not care about their co-op's capacity to play such a role? If that is the case, then perhaps investor-owned firms should be the primary vehicle for the delivery of key public policy initiatives.

The point here is that government renewable energy policy and many of those involved in wind energy development are missing an opportunity to make meaningful contributions toward enhancing community self-governance and development capabilities. Individuals learn democracy in part by doing. However, the emphasis of policy makers and developers in the twentieth century toward rote economic efficiency and command-control management means expert thought leaders have slowly subsumed deliberate debate, dialogue, and collective action. Individuals are increasingly left to feel powerless due to decreasing venues for participation in the "art and science of association." Skills related to problem-solving, building trust, and enhancing reciprocity—and venues for being heard by the powerful—are diminished.

It is troubling that, instead of engagement, communities are left wide open to exploitation through systemic barriers and degradation of self-governing capacities. One cannot count on the federal or state policy mechanisms or on passive development for desired outcomes. However, intentional mobilization through public policy or the development of participatory institutions could go a long way in mitigating the negative aspects of wind energy development and, indeed, enhance the outcomes for collective benefit.

This study was exploratory and could not address all of the critical questions uncovered. What remains to be seen is how the critical consciousness of key leaders emerges, the form it takes, how it is harnessed, and how it is engaged. Do these institutions and their stewards desire to be forces for community well-being against corrosive social forces? Are alternative models to the status quo—specifically, cooperatives—the optimal model for enhanced community economic development? Do cooperatives have the capacity to attempt to become national policy makers by acting in concert with their member-owners from below, not necessarily counting on external agents to take the electric energy system to the next stage? If so, is this a good idea?

A solid conclusion of this research is that there is absolutely no reason why electric cooperatives should be in structural or policy disparity with their IOU counterparts, whether wind or other elements of government energy policy. If energy policy is truly an all-of-the-above proposition, then a variety of institutions must be allowed to take advantage of government subsidy on equal footing with IOUs. Policy, as it stands, privileges absentee, private ownership and stewardship of the grid and natural resources. The general public is, by design, excluded. Should the current trend continue, private, civic-adverse institutions will be the de facto models that proliferate. Democracy is rejected as a matter of policy.

It may seem impossible to broaden public policy to account for electric cooperatives and their role deploying wind energy in an era of government austerity. The capital-intensive nature of wind energy commands significant upfront costs. Capital-intensive projects are going to be out of the reach of rural communities and nonstandard business models (due to state sponsorship of privilege) and will instead suit the private investor. However, this is another problematic area where electric cooperatives offer additional value and solution to policy makers seeking to harness these processes for added community development outcomes.

Energy policy in Denmark and Germany has carved out a role for collective action and crowd-sourced financing, leveraging community buy-in to advance wind energy development. U.S. energy policy could go a long way toward opening the system up to smaller-scale investing by engaging electric cooperatives as leaders in such a process. Electric cooperatives could harness their member-owner network to raise investment capital, the federal government could provide a tax advantage to encourage such investment, and the development of new wind energy infrastructure would be owned by the communities whose natural resources provide the fuel. So if current electric co-op leadership remains complacent, rural Americans have every right to run them out of their positions and replace them with new visionary leadership, to move rural America in a prosperous, forward direction. This seems like a reasonable, low-cost approach and synergistic partnership for energy policy makers to advance wind energy.

There is immense potential for diverse types of institutions outside of "the market" and "the state" to provide solutions to critical social dilemmas facing individuals and their communities. Electric cooperatives—and the cooperative sector as a whole—provide an extraordinarily rich backdrop with which to better understand how to cope with collective action dilemmas and to act upon those solutions to build healthy communities with an engaged citizenry. We must dedicate more energy to understanding this model of the enterprise so that our communities have all of the necessary tools of community economic development at their disposal.

ACES	Alliance for Cooperative Energy Services Power Marketing
AWEA	American Wind Energy Association
CEO	Chief Executive Officer
CFC	National Rural Utilities Cooperative Finance Corporation
CP	ICA cooperative principles and values
DoD	U.S. Department of Defense
DOE	U.S. Department of Energy
FDR	Franklin Delano Roosevelt
FEMA	Federal Emergency Management Agency
FERC	Federal Energy Regulatory Commission
GM	General Manager
G&T	Generation and transmission cooperative
IAD	institutional analysis and development framework
ICA	International Co-operative Alliance
ISO	Independent system operator
IOU	Investor-owned utility
LLC	limited liability company
MACRS	Modified Accelerated Cost Recovery System
MADC	Minot Area Development Corporation
MAFB	Minot Air Force Base
MET tower	Meteorological tower
MISO	Midwest Independent Systems Operators
NERC	North American Electric Reliability Corporation
NRECA	National Rural Electric Cooperative Association
ODP	Ostrom Design Principles
PPA	Power purchasing agreement
PTC	Production tax credit
REA	Rural Electrification Administration
REMC	Rural electric membership cooperative
RPS	renewable portfolio standard
RUS	Rural Utilities Service
USDA	U.S. Department of Agriculture
VEC	Verendrye Electric Cooperative

Introduction

1. For an illustration that rural territories are on the frontlines of wind-rich regions, see image 1, located on my faculty website, https://keith.faculty.ucdavis.edu/.

2. For a chart of wind energy growth and total capacity, see image 2, located on my faculty website, https://keith.faculty.ucdavis.edu/.

3. *IOU* is a standard industry acronym used when referencing investor-owned models of energy utilities.

4. This book focuses on only utility-scale wind farms, as they are the primary recipient of subsidies. Kildegaard and Myers-Kuykindall (2006) classified wind farms with a capacity of fifty megawatts and above as corporate scale. For this book, a wind farm meeting that criterion is more accurately labeled a utility-scale wind farm.

5. An illustration of the spatial scale of the electric co-op sector appears on the official National Rural Electric Cooperative Association (NRECA) website, https://www.electric.coop/we-are-americas-electric-cooperatives/.

6. The primary focus of this research is the community and wind firm interaction, not a complete assessment of government and market actors. For this study, Chapter Four presents a concise analysis of the layers of governance as opposed to complete treatment of federal and state governance.

Chapter One

1. This perspective on free entrance and exit is influenced in part by Tiebout's (1956) "foot-voting model," in which citizens could merely leave as an ultimate expression of dissatisfaction with local institutions of governance.

2. With bounded rationality, individuals pursue goals but do so under constraints of limited cognitive and information-processing capability, incomplete information, and the often subtle influences of cultural predispositions and beliefs (McGinnis, 2011a, p. 173).

3. This is true for most cooperatives. However, utility cooperatives typically service their region in a monopoly capacity; meaning membership is compulsory, not voluntary.

4. Cooperatives operate on a cost-plus basis, meaning they are member-need oriented, not profit oriented. Note that law or institutional governance policy typically caps patronage dividends in order reduce the potential of one member receiving a grossly inordinate amount of benefits above the general membership.

5. For a representation of these three levels of analysis, see image "Three Levels of Analysis," located on my faculty website, https://keith.faculty.ucdavis.edu/.

6. For a detailed illustration of the action arena, see image "Action Arena" located on my faculty website, https://keith.faculty.ucdavis.edu/.

7. While a higher number of participants were interviewed in Illinois, interactions with North Dakota participants were on the whole longer in duration, many of whom spent hours or the entire workday discussing the topic of this research.

8. For a visual representation of the diagnostic elements of IAD, see "IAD," located on my faculty website, https://keith.faculty.ucdavis.edu/.

9. What makes case study design particularly tricky is the lack of resources available on design, and the differentiation of design techniques that are dependent on the phenomenon studied and the "quasi-experimental situation" (Yin, 2008, p. 20). The strength of the IAD framework is that it was crafted through decades of inductive analyses of thousands of case studies, creating a scalable template of sorts that can be consistently applied across many disciplines. This then allows small-N studies to be grouped into large-N meta-analyses.

10. This is represented as "Evaluative Criteria" image located on my website, https://keith.faculty.ucdavis.edu/.

11. The adaptation by Cox, Arnold, and Villamayor Tomas (2010) is favored above Ostrom's original design principles. The Cox et al. adaptation better elaborates on the original through a meta-analysis of the literature. The adaptation validated the design principles and better elaborated upon the original principles.

Chapter Two

1. "Income from certain types of investments qualifies as passive income. Tax paid on this income is considered passive tax. To take advantage of the Federal Production Tax Credit and Modified Accelerated Cost Recovery System, you or a project partner must be paying taxes that fit into this category of tax liability" (Windustry, 2013, n.p.).

2. All personal communications from research interviews and correspondence are described by the generic title of the individual and the date of the communication.

3. EDP Renewables is itself a subsidiary of the Portuguese company Energias de Portugal, S.A. (EDP). In July 2011 Horizon changed its name to EDP Renewables North America LLC.

It should be emphasized that local, national, and global interests drive the wind energy arena. The wind-energy development incentives found in many countries have created a worldwide network of wind energy firms capable of developing and operating wind farms across many national boundaries.

4. Individuals in rural communities throughout Illinois, buoyed by the perception that wind energy development offers economically depressed regions new sources of jobs and revenues, are performing some of the prospecting for wind energy companies on a pro bono basis. McLean County's Center for Renewable Energy (based out of Illinois State University) and the city of Macomb's Illinois Institute for Rural Affair (based out of Western Illi-

nois University) enable these wind prospectors by providing them with a limited number of MET towers to assess local wind resources (university officials, May 10 and 12, 2010).

5. For a representation of the stakeholders, see image "IOU Community Stakeholders" located on my faculty website, https://keith.faculty.ucdavis.edu/.

6. What is often misunderstood is that a local wind farm does not produce electricity for local community consumption (local resident, Sept. 6, 2011; business executive, Dec. 1, 2011). The grid is designed as a spoke-and-wheel system, centralizing power generation in pockets, which then is directed outward toward the grid and end consumers. Once transmitted to the grid, the energy is distributed mechanically, arranged through market transactions or contractual agreements. In virtually all situations, wind energy produced locally is likely transmitted elsewhere regardless of local demand, just as with coal, nuclear, and other energy generation.

7. There are seemingly common misconceptions among consumers related to how the grid itself delivers energy to the end user. The electricity delivered to one's home comes from a mix of sources that cannot be isolated. As one electric cooperative CEO noted: "There isn't such a thing as brown or green electrons. They're all electrons" (May 23, 2012).

8. ComEd's transmission towers were built to accommodate the projected electricity flows. The addition of the wind farms in McLean has added so much unexpected new energy to the lines that it has caused the wires to heat up and sag. ComEd has had to raise the height of the towers (association official, Nov. 11, 2011), part of the unforeseen costs of maintaining transmission infrastructure.

9. The fair market value for electricity in the area was assessed as $360,000 per megawatt. The formula then is $360,000 per megawatt × 1.65 megawatts per turbine × 240 total turbines = $142,560,000 fair market value for the Horizon wind farm. Each 1.65-megawatt turbine was assessed at a value of $594,000 or, in other words, was taxed as if it were a house with a valuation of $594,000. Each turbine also gets an annual depreciation of 4 percent over twenty-five years, which is applied toward reducing the wind company's tax liability, meaning the value of the turbines will decrease as well (school official, Nov. 11, 2001).

Chapter Three

1. Since the completion of PrairieWinds ND 1, Inc. in 2009, the parent cooperative, Basin Electric Power Cooperative, has opened the nation's second cooperative wind farm, in South Dakota.

2. My own car broke down on an isolated road outside of Bismarck, North Dakota. I can report anecdotally from my own experience that thankfully there is an element of truth to this statement.

3. The phenomenon of increasing rural property values is not restricted to Ward County, North Dakota, or other regions experiencing a rush on oil and natural gas assets. Speculation on land also impacts rural regions with rich agriculture due to the increasing growth in commodity agriculture and biofuel development.

4. The crisis atmosphere made it difficult to interview key public officials.

5. For a figure representing the Basic Electric Power Cooperative, see image "Basin" located on my faculty website, https://keith.faculty.ucdavis.edu/.

6. Peaking plants are power plants meant to rapidly deploy energy to the grid at peak usage, when standard baseload electricity levels are not meeting projected demand.

7. For a figure representing the Basic Electric Power Cooperative, see "Basin 2" located on my faculty website https://keith.faculty.ucdavis.edu/.

Chapter Four

1. For an illustration of the electric grid, see "grid image." located on my faculty website, https://keith.faculty.ucdavis.edu/.

2. Critics use the intermittency of wind energy to claim that wind energy is far more unpredictable than other energy sources. One must also consider that if transportation facilities shut down, regulations hamper mining and drilling, or nonrenewables become scarce, they too are in a sense intermittent. What makes them stand out from wind is their stable physical form, which allows them to sit in storage during periods of surplus accumulation, stockpiled like batteries until needed. Wind energy has yet to enjoy the development of an affordable, scalable battery system that would allow stored surplus energy stocks to flow into the grid as needed.

3. For a complete list of wind energy subsidies, see Climate Policy Initiative (2012, p. 9).

4. The central wind energy association, American Wind Energy Association (AWEA), claims the subsidy has been a resounding success. AWEA is calling for a phase-out of federal subsidies due in part to projections that wind energy will be price-competitive with subsidized fossil fuels in under a decade.

5. The ODPs have recently been extended outside the realm of natural resource governance. Ostrom's colleagues at the Bloomington School have applied the ODPs to examine the health care commons, and Ostrom has a book on knowledge as a commons. The application beyond natural resource management is justified as such:

> Over several years she [Ostrom] examined case studies of natural resource commons which were successfully managed by local users of that commons over long periods of time, as well as cases in which these efforts were unsuccessful. Community groups exert stewardship by establishing and enforcing their own rules concerning how many and what types of resources can be extracted, and when, as well as requiring contributions to collective efforts to maintain access to those resources. Ostrom summarizes her findings in an influential list of eight "design principles" which are satisfied, in one way or another, in cases of sustainable resource management. However, Ostrom's work on sustainable resource management had been tightly linked to natural resources, and not yet widely applied to standard businesses, or to highly technological systems. Our initial discussions focused on the serious concern that lessons drawn from the study of natural resource management (mostly in the developing world) might not even be relevant to the highly

technical realm of modern healthcare. Technically speaking, Ostrom limited her con-
clusions to the management of common-pool resources, in which individuals extract
resources from a common pool for their use. Some commons are better described as
public goods, in the sense that individuals jointly enjoy the benefits without any threat
of exhaustion. Other commons are available only to those who pay a membership fee,
as is the case for country clubs or housing associations. Health and healthcare policy
encompasses the full array of private, public, and club goods, and only a few aspects fit
the technical definition of a common pool resource. (McGinnis & Brink, 2012, pp. 1–2)

McGinnis goes on to demonstrate how a complex system (health care) comprising a mix
of goods can be conceptualized as a commons. Access to emergency department services
seems the best fit since emergency departments are subject to overcrowding and overuse in
some circumstances. The patients are users of an emergency department who need to draw
on the skills of the physicians and nurses to improve their health. There are a limited number
of medical personnel who can treat a finite amount of patients at one time, just as there are a
limited number of examination areas. If a patient comes to the emergency department for a
nonemergency, the doctor who treats that patient is not able to take care of another patient
who needs emergency care.

Other important aspects of health care, like community health or insurance coverage, are
more like public or club goods. Fortunately, Ostrom left open the possibility that the design
principles might also be relevant to the sustainable production of public goods, especially
at the local level (p. 2).

Chapter Five

1. For an illustration of how policy process reinforces privilege of the investor-owned
firm while relegating the community to a passive participant, see image "Policy Impacts on
Utility Types," located on my faculty website, https://keith.faculty.ucdavis.edu/.

2. For an illustration of how co-ops and their communities could apply upward pressure
on other institutions, see image "Policy from Below," located on my faculty website, https://
keith.faculty.ucdavis.edu/.

REFERENCES

ACES (Alliance for Cooperative Energy Services Power Marketing). (2016). About. Retrieved September 15, 2016, from http://www.acespower.com/about/

Adamson, M. R. (2008). Oil booms and boosterism: Local elites, outside companies, and the growth of Ventura, California. *Journal of Urban History, 35*, 150–177. doi:10.1177/0096144208321873

Aitken, M. (2010). Why we still don't understand the social aspects of wind power: A critique of key assumptions within the literature. *Energy Policy, 38*, 1834–1841. doi:10.1016/j.enpol.2009.11.060

Albrecht, S. L., Finsterbusch, K., Freudenburg, W. R., Gale, R. P., Gold, R. L., Murdock, S. H., & Leistritz, F. L. (1982). Commentary on "Local social disruption and western energy development: A critical review." *Pacific Sociological Review, 25*(3), 297–366.

Aligica, P. D., & Boettke, P. (2009). *Challenging institutional analysis and development: The Bloomington School*. New York, New York: Routledge.

Aligica, P. D., & Tarko, V. (2012). Polycentricity: From Polanyi to Ostrom, and beyond. *Governance, 25*, 237–262. doi:10.1111/j.1468-0491.2011.01550.x

Alinsky, S. D. (1971). *Rules for radicals*. New York, New York: Random House.

Alperovitz, G. (2011). *American beyond capitalism: Reclaiming our wealth, our liberty, and our democracy* (2nd ed.). Hoboken, New Jersey: Wiley.

Anderson, C. (2006). Farms want to reap the wind. *Pantagraph*. Retrieved May 15, 2018, from http://www.pantagraph.com

AWEA (American Wind Energy Association). (2012a). The American wind industry urges Congress to take immediate action to pass an extension of the PTC [Press release]. Retrieved September 14, 2016, from http://www.awea.org/issues/federal_policy/upload/PTC-Fact-Sheet.pdf

AWEA (American Wind Energy Association). (2012b). *AWEA U.S. Wind Industry Fourth Quarter 2012 Market Report*. Retrieved September 14, 2016, from http://www.awea.org/learnabout/publications/reports/upload/AWEA-Fourth-Quarter-Wind-Energy-Industry-Market-Report_Executive-Summary-4.pdf

AWEA (American Wind Energy Association). (2016). *Market reports: Ownership and rankings—Issues*. Retrieved September 14, 2016, from http://www.awea.org/AnnualMarketReport.aspx?ItemNumber=6311

Bacigalupi, L., & Freudenberg, W. (1983). Increased mental health caseloads in an energy boomtown. *Administration in Mental Health, 10*, 306–322. doi:10.1007/BF00823107

Bäckstrand, K. (2006). Democratizing global environmental governance? Stakeholder democracy after the world summit on sustainable development. *European Journal of International Relations, 12*(4), 467–498.

Bailey, D. (2011). Race on to house Minot flood-displaced before winter. Reuters.

Retrieved May 15, 2018, from https://www.reuters.com/article/us-northdakota-flooding/race-on-to-house-minot-flood-displaced-before-winter-idUSTRE79D61X20111014

Basin Electric Power Cooperative. (2009). Minot wind project in full operation. [Press release]. Retrieved March 1, 2013, from http://www.basinelectric.com/News_Center/Publications/News_Releases/Minot_wid_project_in_full_operation.html

Basin Electric Power Cooperative. (2012). Cooperatives work together to find housing solutions in the Bakken. Basin Electric News Center. Retrieved September 14, 2016, from http://www.basinelectric.com/News_Center/Publications/News_Briefs/cooperatives-work-together-to-find-housing-solutions-in-the-Bakken.html

Basin Electric Power Cooperative. (2016a). 1958–1960. Retrieved September 15, 2016, from https://www.basinelectric.com/About-Us/Organization/History/1958-1960/

Basin Electric Power Cooperative. (2016b). 1961–1984. Retrieved September 15, 2016, from https://www.basinelectric.com/About-Us/Organization/History/1961-1984/

Basin Electric Power Cooperative. (2016c). About us. Retrieved September 15, 2016, from https://www.basinelectric.com/About-Us/

Basin Electric Power Cooperative. (2016d). At a glance. Retrieved September 15, 2016, from https://www.basinelectric.com/About-Us/Organization/At-a-Glance/

Basin Electric Power Cooperative. (2016e). Governance model. Retrieved September 15, 2016, from https://www.basinelectric.com/About-Us/Organization/Governance/

Basin Electric Power Cooperative. (2016f). Governance model. Retrieved May 15, 2018, from http://www.basinelectric.com/About_Us/Corporate/Governance_Model/index.html

Basin Electric Power Cooperative. (2016g). History. Retrieved September 15, 2016, from https://www.basinelectric.com/About-Us/Organization/History/index.html

Bauman, Z. (2001). *Community: Seeking safety in an insecure world.* Cambridge, England: Polity Press.

Bawden, T. (2012). Davos call for $14trn "greening" of global economy. *The Independent.* Retrieved May 15, 2018, from https://www.independent.co.uk/news/business/news/davos-call-for-14trn-greening-of-global-economy-8460994.html

Becht, M., Bolton, P., & Röell, A. (2003). Corporate governance and control. *Handbook of the Economics of Finance, 1,* 1–109.

Bettenhausen, T. (2011). The art of the possible: A look into the future via the history of Dakota Gas. *Basin Today Magazine,* (4), 2–5.

Black, D., McKinnish, T., & Sanders, S. (2005). The economic impact of the coal boom and bust. *The Economic Journal, 115,* 449–476. doi:10.1111/j.1468-0297.2005.00996.x

Blomkvist, P., & Larsson, J. (2013). An analytical framework for common-pool resource –large technical system (CPR-LTS) constellations. *International Journal of the Commons, 7*(1), 113–139.

Bloomington-Normal Economic Development Council. (2013). Demographic profile [Fact sheet]. Retrieved April 5, 2017, from http://www.bnbiz.org/datacenter/demographic_profile.php

Borzaga, C., & Galera, G. (2012). *Promoting the understanding of cooperatives for a better world.* Retrieved May 15, 2018, from http://www.euricse.eu/node/1736

Bowles, S., & Gintis, H. (2002). Social capital and community governance. *The Economic Journal, 112*, F419–F436. doi:10.1111/1468-0297.00077

Brady-Lunny, E. (2007). Court to hear wind farm battle. *Pantagraph*. Retrieved August 5, 2011, from http://www.pantagraph.com/articles/2007/07/24/news/doc46a6b48258 a18719179882.txt

Brennan, M. A. (2009). Cooperatives as tools for community and economic development in Florida (no. FCS9208). IFAS Community Development Series. Retrieved April 5, 2017, from http://edis.ifas.ufl.edu/

Broadway, M. J., & Stull, D. D. (2006). Meat processing and Garden City, KS: Boom and bust. *Journal of Rural Studies, 22*(1), 55–66. doi:10.1016/j.jrurstud.2005.06.001

Brockhouse, B. (2008). *Lassoing Wyoming's wind: Landowners band together to bargain for higher wind-power royalties*. Retrieved April 5, 2017, from http://www.rurdev.usda.gov /supportdocuments/CoopMag-jul08.pdf

Brown, R. B., Dorins, S. F., & Krannich, R. S. (2005). The boom-bust-recovery cycle: Dynamics of change in community satisfaction and social integration in Delta, Utah. *Rural Sociology, 70*(1), 28–49.

Brown, S. V., Nderitu, D. G., Preckel, P. V., Gotham, D. J., & Allen, B. W. (2011). *Renewable Power Opportunities for Rural Communities*. U.S. Department of Agriculture, Office of Chief Economist. Retrieved May 15, 2018, from http://www.usda.gov/oce/reports /energy/RenewablePowerOpportunities-Final.pdf

Burt, R. S. (1995). *Structural holes: The social structure of competition*. Cambridge, Massachusetts: Harvard University Press.

Carson, K. A. (2010). The distorting effects of transportation subsidies. *The Freeman*. Retrieved October 4, 2018, from https://fee.org/articles/the-distorting-effects -of-transportation-subsidies/

Carter, J. (2011). *The effect of wind farms on residential property values in Lee County, Illinois*. Retrieved May 5, 2018, from http://renewableenergy.illinoisstate.edu/downloads /publications/2011%20Wind%20Farms%20Effect%20on%20Property%20Values %20in%20Lee%20County.pdf

Cracogna, D., Fici, A., & Henrÿ, H. (Eds.). (2013). *International handbook of cooperative law*. Berlin: Springer.

Center for Renewable Energy. (2013). Retrieved September 16, 2016, from http:// renewableenergy.illinoisstate.edu/

Christakis, N. A., & Fowler, J. H. (2009). *Connected: The surprising power of our social networks and how they shape our lives*. Boston, Massachusetts: Little, Brown.

Climate Policy Initiative. (2012). Supporting renewables while saving taxpayers money. Retrieved October 4, 2018, from http://climatepolicyinitiative.org/wp-content /uploads/2012/09/Supporting-Renewables-while-Saving-Taxpayers-Money.pdf

Cohan, D. (2016). When coal companies go bankrupt, the mining doesn't always stop. *The Hill*. Retrieved May 5, 2018, from http://thehill.com/blogs/pundits-blog /energy-environment/276628-when-coal-companies-go-bankrupt-the-mining-doesnt

Cook, M. L. (1994). The role of management behavior in agricultural cooperatives. *Journal of Agricultural Cooperation, 9*, 42–58.

Cooper, J. (2008). Electric co-operatives: From New Deal to bad deal. *Harvard Journal on Legislation, 45*, 335–375.

Coulter, P. (2006). Governor announces grant for McLean Co. wind farm. *Pantagraph.* Retrieved October 4, 2018, from https://www.pantagraph.com/news/governor -announces-grant-for-mclean-co-wind-farm/article_8aaa4876-9a26-5c66-8bd4 -98012903a61a.amp.html

Coulter, P. (2009). School districts see financial safety net in wind farms. *Pantagraph.* Retrieved October 4, 2018, from https://www.pantagraph.com/news/local/school -districts-see-financial-safety-net-in-wind-farms/article_0ab662be-6cd1-11de-a32e -001cc4c002e0.html

Cox, M., Arnold, G., & Villamayor Tomas, S. (2010). A review of design principles for community-based natural resource management. *Ecology and Society, 15*(4), 1–19.

Creswell, J. W. (2008). *Research design: Qualitative, quantitative, and mixed methods approaches.* Los Angeles, California: Sage.

Cunningham, T. H. (2011). Lengthy flood recovery seen in N.D. *Electric Co-op Today.* Retrieved May 5, 2018, from http://www.ect.coop

Davis, K. S. (1986). *FDR: The New Deal years 1933–1937.* New York, New York: Random House.

Deller, S., Hoyt, A., Hueth, B., & Sundaram-Stukel, R. (2009). *Research on the economic impact of cooperatives.* Retrieved May 5, 2018, from http://reic.uwcc.wisc.edu/

DeMeo, E. (2003). *Some common misconceptions about wind power.* Retrieved October 4, 2018, from http://infohouse.p2ric.org/ref/45/44525.pdf

Duménil, G., & Lévy, D. (2004). *Capital resurgent: Roots of the neoliberal revolution.* London, England: Harvard University Press.

Eisenhardt, K. M., & Graebner, M. E. (2007). Theory building from cases: Opportunities and challenges. *Academy of Management Journal, 50*(1), 25–32. doi:10.5465/AMJ.2007 .24160888

Emery, M., & Flora, C. (2006). Spiraling-up: Mapping community transformation with Community Capitals Framework. *Community Development: Journal of the Community Development Society, 37*(1), 19–35. doi:10.1080/15575330609490152

Evans, P. (2004). Development as institutional change: The pitfalls of monocropping and the potentials of deliberation. *Studies in Comparative International Development, 38*(4), 30–52. doi:10.1007/BF02686327

Fazzi, L. (2011). Social co-operatives and social farming in Italy. *Sociologia Ruralis, 51*, 119–136. doi:10.1111/j.1467-9523.2010.00526.x

Finzel, B., & Kildegaard, A. (2013). *Whose wind? Prospects for cooperative and community wind development on the U.S. Upper Great Plains.* Manuscript submitted for publication.

Flint, C. G., Luloff, A. E., & Theodori, G. L. (2010). Extending the concept of community interaction to explore regional community fields. *Journal of Rural Social Sciences, 25*(1), 22–36.

Flora, C. B., & Flora, J. L. (2008). *Rural communities: Legacy and change* (3rd ed.). Boulder, Colorado: Westview Press.

Florini, A., & Sovacool, B. K. (2009). Who governs energy? The challenges facing global energy governance. *Energy Policy, 37,* 5239–5248. doi:10.1016/j.enpol.2009.07.039

Ford, M. A. (2006a). Planned wind farm to include visitor's center. *Pantagraph.* Retrieved August 5, 2011, from http://www.pantagraph.com

Ford, M. A. (2006b). Wind farm to file for permit Wednesday. *Pantagraph.* Retrieved October 4, 2018, from https://www.pantagraph.com/news/wind-farm-to-file-for-permit -wednesday/article_f8e67e84-97ff-57e2-8f2b-ba519b156337.html

Ford, M. A. (2010). Board to hold hearing on proposed wind farm. *Pantagraph.* Retrieved October 4, 2018, from https://www.pantagraph.com/news/local/board-to-hold-hearing -on-proposed-wind-farm/article_5612a63e-a80b-11df-ab30-001cc4c002e0.html

Freeman, L. C. (2004). *The development of social network analysis: A study in the sociology of science.* Vancouver, British Columbia: Empirical Press.

FreeState Electric Cooperative. (2016). Touchstone Energy. Retrieved September 15, 2016, from http://www.ljec.coopwebbuilder.com/content touchstone-energy

Freudenberg, W. (1979). *People in the impact zone: The human and social consequences of energy boomtown growth in four Western Colorado communities.* Unpublished doctoral dissertation. Ann Arbor, Michigan: University of Michigan Microfilms.

Freudenburg, W. R. (1984). Boomtown's youth: The differential impacts of rapid community growth on adolescents and adults. *American Sociological Review, 49*(5), 697–705.

Freudenberg, W.R. (1986). The density of acquaintanceship: An overlooked variable in community research. *American Journal of Sociology, 92,* 27-63.

Freudenburg, W. R. (2008). Thirty years of scholarship and science on environment-society relationships. *Organization and Environment, 21,* 449–459.

Fundingsland, K. (2012, March 4). Calling all contractors: Plenty of work remains as Minot rebuilds. *Minot Daily News.*

Gaventa, J. (1982). *Power and powerlessness: Quiescence and rebellion in an Appalachian valley.* Urbana, Illinois: University of Illinois Press.

Gaventa, J. (2002). Exploring citizenship, participation and accountability. *IDS Bulletin, 33*(2), 1–14.

Glans, M. (2012). Subsidies for renewable energy in Florida. The Heartland Institute. Retrieved May 5, 2018, from http://heartland.org/policy-documents/research -commentary-subsidies-renewable-energy-florida

Ghoshal, S. (2005). Bad management theories are destroying good management practices. *Academy of Management Learning and Education, 4*(1), 75–91.

Granovetter, M. S. (1973). The strength of weak ties. *American Journal of Sociology, 78,* 1360–1380. doi:10.1086/225469

Greer, M. L. (2008). A test of vertical economies for nonvertically integrated firms: The case of rural electric cooperatives. *Energy Economics, 30,* 679–687. doi:10.1016/j .eneco.2006.08.001

Hansmann, H. (2000). *The ownership of enterprise.* Harvard University Press.

Holliday, B. (2009). New ISU study: Wind energy may generate $2 billion in benefits in Illinois. *Pantagraph.* Retrieved October 4, 2018, from https://www.pantagraph.com

/business/new-isu-study-wind-energy-may-generate-billion-in-benefits/article
_002b6a84-7194-11de-87b9-001cc4c002e0.html

Holly, D. (2011). North Dakota flooding continues. *Electric Co-op Today*. Retrieved December 2, 2011, from http://www.ect.coop

Holly, D. (2012). N.D. co-ops come to aid of Minot. *Electric Co-op Today*. Retrieved May 16, 2013, from http://www.ect.coop

Hughes, T. P. (1976). Technology and public policy: The failure of giant power. *Proceedings of the IEEE, 64*(9), 1361–1371. doi:10.1109/PROC.1976.10327

Illinois Department of Commerce and Economic Opportunity. (2016a). Tax assistance: High impact business (HIB). Retrieved September 16, 2016, from https://www.illinois.gov/dceo/expandrelocate/incentives/taxassistance/pages/hib.aspx

Illinois Department of Commerce and Economic Opportunity. (2016b). Tax assistance: Illinois Enterprise Zone Program. Retrieved September 15, 2016, from https://www.illinois.gov/dceo/expandrelocate/incentives/taxassistance/pages/enterprisezone.aspx

International Association for Community Development. (2012). About. Retrieved September 6, 2016, from http://www.iacdglobal.org/about

International Co-operative Alliance. (2015). Retrieved March 29, 2015, from http://ica.coop/en/what-co-op/co-operative-identity-values-principles

Isserman, A. M., Feser, E., & Warren, D. E. (2009). Why some rural places prosper and others do not. *International Regional Science Review, 32*, 300–342.

Jesover, F., & Kirkpatrick, G. (2005). The revised OECD principles of corporate governance and their relevance to non-OECD countries. *Corporate Governance: An International Review, 13*(2), 127–136.

Johnson, R. (2012). You've never seen anything like this North Dakota oil boomtown. *Business Insider*. Retrieved October 4, 2018, from https://www.businessinsider.com.au/youve-never-seen-anything-like-the-williston-oil-boom-2012-3#williston-north-dakota-is-in-the-northwestern-portion-of-the-state-not-far-from-montana-and-canada-1

Kildegaard, A., & Myers-Kuykindall, J. (2006). *Community wind versus corporate wind: Does it matter who develops the wind in Big Stone County, MN?* Retrieved October 4, 2018, from http://www.windustry.org/resources/community-vs-corporate-wind-does-it-matter-who-develops-wind-big-stone-county-mn

Ki-moon, B. (2009). *Cooperatives in social development: Report of the secretary-general* (no. A/64/132). Retrieved May 5, 2018, from http://daccess-dds-ny.un.org/doc/UNDOC/GEN/N09/402/01/PDF/N0940201.pdf?OpenElement

Klein, N. (2007). *The shock doctrine: The rise of disaster capitalism*. New York, NY: Macmillan.

Knoke, D., & Yang, S. (2008). *Social network analysis* (2nd ed.). Thousand Oaks, California: Sage.

Kolpack, D. (2011). In Minot, focus on protecting critical services. ABC News. Retrieved May 16, 2013, from http://abcnews.go.com

Laloux, F. (2015). *Reinventing organizations*. Belgium: Nelson Parker.

Ledwith, M., & Campling, J. (2005). *Community development: A critical approach* (2nd ed.). Bristol, United Kingdom: Policy Press.

Loomis, D., & Aldeman, M. (2011). *Wind farm implications for school district revenue.* Retrieved May 16, 2013, from http://renewableenergy.illinoisstate.edu/downloads/publications/2011%20School%20District%20Report.pdf

Loomis, D., & Carter, J. (2011). *Economic impact: Wind energy development in Illinois.* Retrieved May 16, 2013, from http://renewableenergy.illinoisstate.edu/downloads/publications/2012EconomicImpactReportForWeb.pdf

Lowery, M. (2010). On the shoulders of giants. *Management Quarterly, 51*(1), 4–14.

Lukes, S. (2004). *Power: A radical view* (2nd ed.). London, England: Palgrave Macmillan.

Madison, C. (2010). For wind, 2009 was a boom year, but we need RES for jobs [Blog post]. Retrieved May 16, 2013, from http://www.awea.org/blog/index.php?mode=viewid&post_id=302

Martinez, C. (2009). Barriers and challenges of implementing tobacco control policies in hospitals: Applying the Institutional Analysis and Development Framework to the Catalan network of smoke-free hospitals. *Policy, Politics, and Nursing Practice, 10*, 224–232. doi:10.1177/1527154409346736

McGinnis, M. D. (2011a). An introduction to IAD and the language of the Ostrom Workshop: A simple guide to a complex framework. *Policy Studies Journal, 39*, 169–183. doi:10.1111/j.1541-0072.2010.00401.x

McGinnis, M. D. (2011b). Networks of adjacent action situations in polycentric governance. *Policy Studies Journal, 39*, 51–78. doi:10.1111/j.1541-0072.2010.00396.x

McGinnis, M. D., & Brink, C. A. (2012). *Shared stewardship of a health commons: Examples and opportunities from Grand Junction, Colorado.* Alliance of Community Health Plans. Retrieved October 4, 2018, from http://php.indiana.edu/~mcginnis/healthcommons/HealthCommonsWhitePaper.pdf

McKibben, B. (2007). *Deep economy: The wealth of communities and the durable future.* New York, New York: Macmillan.

Melillo, J. M., Richmond, T. C., and Yohe, G. W. (Eds.). (2014). *Climate change impacts in the United States: The Third National Climate Assessment.* U.S. Global Change Research Program. doi:10.7930/J0Z31WJ2

Miller, S. (2006a). FAA may have stopped wind farm work. *Pantagraph.* Retrieved October 4, 2018, from https://www.pantagraph.com/news/faa-may-have-stopped-wind-farm-work/article_19222e39-d32d-5a47-adc3-3ede3492f29a.html

Miller, S. (2006b). Wind farm construction to begin. *Pantagraph.* Retrieved August 5, 2011, from http://www.pantagraph.com

Miller, S. (2006c). Wind farm on hold for wetland conservation. *Pantagraph.* Retrieved October 4, 2018, from https://www.pantagraph.com/news/wind-farm-on-hold-for-wetland-conservation/article_5a5864b1-c6d8-5ef3-9201-dad702c06c4e.html

Miller, S. (2007a). Central Illinois "hotbed" for wind energy. *Pantagraph.* Retrieved October 4, 2018, from https://www.pantagraph.com/business/central-illinois-hotbed-for-wind-energy/article_16c0c831-93cf-5c92-9f38-8d2a66c51cde.html

Miller, S. (2007b). Twin Groves wind farm may add 170 towers. *Pantagraph*. Retrieved October 4, 2018, from https://www.pantagraph.com/news/twin-groves-wind-farm -may-add-towers/article_c651f39f-2718-53f9-baf3-fcc090f40f74.html

Minot Area Chamber of Commerce. (2013). Population and employment [Fact sheet]. Retrieved March 1, 2013, from http://www.minotchamber.org/community/popemploy .shtml

Minot Area Development Corporation. (2013a, January 15). *Job Service North Dakota*. Retrieved October 4, 2018, from http://www.minotusa.com/uploads%5Cresources %5C118%5C2012-year-end-economic-indicators.pdf

Minot Area Development Corporation. (2013b). About: Board members [Fact sheet]. Retrieved March 1, 2013, from http://www.minotusa.com/about/board-of-directors/

Minot Area Development Corporation. (2013c). Facts and figures [Fact sheet]. Retrieved March 1, 2013, from http://www.minotusa.com/factsandfigures/

Minot Daily News. (2012a). Federal infrastructure aid for 2011 flood tops $200M. Retrieved March 1, 2013, from http://www.minotdailynews.com

Minot Daily News. (2012b). Proposed oil and gas rule changes now final. Retrieved March 1, 2013, from http://www.minotdailynews.com

Minot Recovery Information. (2013). 2011 flood facts [Fact sheet]. Retrieved March 1, 2013, from http://www.minotrecoveryinfo.com/facts/2011-flood-by-the-numbers/

Miraftab, F. (2004). Public-private partnerships: The Trojan horse of neoliberal development? *Journal of Planning Education and Research, 24*, 89–101. doi:10.1177 /0739456X04267173

Mooney, P. H. (2004). Democratizing rural economy: Institutional friction, sustainable struggle and the cooperative movement. *Rural Sociology, 69*, 76–98. doi:10.1526 /003601104322919919

Murray, C. (2011). *Co-op survival rates in British Columbia*. Retrieved March 1, 2013, from http://www.bcca.coop/sites/bcca.coop/files/BALTA_A11_report_BC.pdf

National Rural Utilities Cooperative Finance Corporation. (2013a). Cooperative work-place and workforce [Fact sheet]. Retrieved March 1, 2013, from http://www.nreca .coop/issues/CoopWorkplaceWorkforce/Pages/default.aspx

National Rural Utilities Cooperative Finance Corporation. (2013b). Overview [Fact sheet]. Retrieved March 1, 2013, from https://www.nrucfc.coop/content/cfc/about _cfc/overview.html

N.C. Clean Energy Technology Center. (2012). DSIRE—Database of State Incentives for Renewables and Efficiency: Minnesota renewables portfolio standard [Fact sheet]. Retrieved March 1, 2013, from http://www.dsireusa.org/incentives/incentive .cfm?Incentive_Code=MN14R&re=1&ee=0

North Dakota Industrial Commission. (2016). Renewable Energy Program list of grant projects. Retrieved September 16, 2016, from http://www.nd.gov/ndic/renew-project .htm

North Dakota Transmission Authority. (2009). *Annual report: July 1, 2008 to June 30, 2009*. Retrieved August 5, 2011, from www.nd.gov/ndic/ic-press/ta-annualreport-09.pdf

NRECA International. (2016). *Guides for electric cooperative development and rural electrification.* Retrieved September 1, 2016, from https://www.nreca.coop/wp-content/uploads/2013/07/GuidesforDevelopment.pdf

NRECA Twenty-First Century Cooperative Committee. (2013). *The electric cooperative purpose: A compass for the 21st century (report of the NRECA 21st Century Cooperative Committee).* Arlington, Virginia: National Rural Electric Cooperative Association.

Oil Change International. (2013). Fossil fuel subsidies in the U.S. [Fact sheet]. Retrieved March 1, 2013, from http://priceofoil.org/fossil-fuel-subsidies/

Ondracek, J. (2011). *Minot, North Dakota housing demand analysis.* Retrieved March 1, 2013, from http://www.minotusa.com/uploads/resources/87/minot-housing-study-exec-summary.pdf

Ostrom, E. (1990). *Governing the commons: The evolution of institutions for collective action.* New York, New York: Cambridge University Press.

Ostrom, E. (1999). Coping with tragedies of the commons. *Annual Review of Political Science, 2,* 493–535. doi:10.1146/annurev.polisci.2.1.493

Ostrom, E. (2005). *Understanding institutional diversity.* Princeton, New Jersey: Princeton University Press.

Ostrom, E. (2009). *A polycentric approach for coping with climate change.* The World Bank.

Ostrom, E. (2010). A long polycentric journey. *Annual Review of Political Science, 13,* 1–23. doi:10.1146/annurev.polisci.090808.123259

Ostrom, E. (2014). A polycentric approach for coping with climate change. *Annals of Economics and Finance, 15*(1), 97–134.

Ostrom, E., & Kiser, L. (2000). The three worlds of action: A meta-theoretical synthesis of institutional approaches. In M. McGinnis, *Polycentric games and institutions: Readings from the Workshop in Political Theory and Policy Analysis* (pp. 56–88). Ann Arbor, Michigan: University of Michigan Press.

Ostrom, V. (1989). *The intellectual crisis in American public administration* (2nd ed.). Tuscaloosa, Alabama: University of Alabama Press.

Ostrom, V. (1997). *The meaning of democracy and the vulnerability of democracies: A response to Tocqueville's challenge.* Ann Arbor, Michigan: University of Michigan Press.

Pantagraph. (2006). Durbin, Obama criticize wind farm orders. Retrieved October 4, 2018, from https://www.pantagraph.com/news/durbin-obama-criticize-wind-farm-orders/article_db2733d5-b2ee-5f64-a978-f120d088946b.html

Pantagraph. (2008). Exelon purchases power from county's Twin Groves wind farm. Retrieved October 4, 2018, from https://www.pantagraph.com/business/exelon-purchases-power-from-county-s-twin-groves-wind-farm/article_c5a2d57a-7432-5d4a-8a28-2a2125b34a94.html

Patrie, B. (1998). *Creating "co-op fever"—A rural developers guide to developing cooperatives.* Retrieved from http://www.rurdev.usda.gov/rbs/pub/sr54.pdf

Pellow, D. N., & Brulle, R. J. (2005). *Power, justice, and the environment: A critical appraisal of the environmental justice movement.* Cambridge, Massachusetts: MIT Press.

Poteete, A. R., Janssen, M. A., & Ostrom, E. (2010). *Working together: Collective action, the commons, and multiple methods in practice.* Princeton, New Jersey: Princeton University Press.

Preskey, D. (2012). ND crime rate up. KXNet. Retrieved March 1, 2013, from http://www.kxnet.com

Putnam, R. D. (2000). *Bowling alone: The collapse and revival of American communities.* New York, New York: Simon and Shuster.

Putnam, R. D., Leonardi, R., & Nanetti, R. Y. (1993). *Making democracy work: Civic traditions in modern Italy.* Princeton, New Jersey: Princeton University Press.

Restakis, J. (2010). *Humanizing the economy: Co-operatives in the age of capital.* Gabriola Island, British Columbia: New Society Publishers.

Riopell, M. (2009). Can crop dusters and wind farms coexist? *Pantagraph.* Retrieved October 4, 2018, from https://www.pantagraph.com/business/can-crop-dusters-and-wind-farms-coexist/article_69022768-e9fe-56b1-a99e-9fae30ac20dd.html

Rostow, W. W. (1960). *The stages of economic growth: A noncommunist manifesto.* Cambridge, England: Cambridge University Press.

Santos, F. M. (2012). A positive theory of social entrepreneurship. *Journal of Business Ethics, 111,* 335–351. doi:10.1007/s10551-012-1413-4

Sapochetti, T. (2010). Wind farm developments power up in Central Illinois. *Pantagraph.* Retrieved October 4, 2018, from https://www.pantagraph.com/business/wind-farm-developments-power-up-in-central-illinois/article_4891b82c-3850-11df-87b1-001cc4c03286.html

Schramm, J. (2010). Powering up: Oil boom energizes Mountrail-Williams Electric. *Minot Daily News.* Retrieved March 1, 2013, from http://www.minotdailynews.com

Schramm, J. (2012). Micropolitan magic: Minot region makes census list of fastest-growing areas. *Minot Daily News.* Retrieved March 1, 2013, from http://www.minotdailynews.com

Scott, J. (2000). *Social network analysis: A handbook* (2nd ed.). London, England: Sage.

Scott, J. C. (1999). *Seeing like a state: How certain schemes to improve the human condition have failed.* Binghamton, New York: Vali-Ballou Press.

Sen, A. (2011). *The idea of justice.* Cambridge, Massachusetts: Harvard University Press.

Sharp, J. S. (2001). Locating the community field: A study of interorganizational network structure and capacity for community action. *Rural Sociology, 66,* 403–424. doi:10.1111/j.1549-0831.2001.tb00074.x

Sherwood, A., & Taylor, K. (2014). Unique expectations of cooperative boards: Taking on the challenges of the democratic enterprise. *International Journal of Cooperative Management, 7*(1), 29–42.

Shragge, E., & Toye, M. (2006). *Community economic development: Building for social change.* Nova Scotia: Cape Breton University Press.

Shults, J. W. (2006). Eighty-turbine wind farm coming to Benson. *Pantagraph.* Retrieved March 1, 2013, from http://www.pantagraph.com

Small, M. L. (2009). *Unanticipated gains: Origins of network inequality in everyday life.* New York, New York: Oxford University Press.

Spillane, J. P. (2012). *Distributed leadership* (Vol. 4). San Francisco, California: Wiley.

Stagl, J. (2012). Railroad coal traffic update: Domestic volumes are down, but demand for export coal is up. Progressive Railroading. Retrieved October 4, 2018, from https://www.progressiverailroading.com/rail_industry_trends/article/Railroad-coal-traffic-update-Domestic-volumes-are-down-but-demand-for-export-coal-is-up--30944

Stark, N. (2007). *Eight principles for effective rural governance and how communities put them into practice.* Retrieved August 5, 2011, from Center for Rural Entrepreneurship website: http://www.rupri.org/Forms/RGIreport.pdf

Stedman, R., Lee, B., Brasier, K., Weigle, J. L., & Higdon, F. (2009). Cleaning up water? Or building rural community? Community watershed organizations in Pennsylvania. *Rural Sociology, 74,* 178–200. doi:10.1111/j.1549-0831.2009.tb00388.x

Steever, T. (2011). McLean County is Illinois's biggest corn, soybean producer. *Brownfield: Ag News for America.* Retrieved March 1, 2013, from http://brownfieldagnews.com

Steinberg, D., & Porro, G. (2012). *Preliminary analysis of the jobs and economic impacts of renewable energy projects supported by the §1603 Treasury grant program.* National Renewable Energy Laboratory technical no. NREL/TP-6A20-52739. Retrieved March 1, 2013, from www.nrel.gov/docs/fy12osti/52739.pdf

Stoel Rives, LLP. (2010). *The law of wind: A guide to business and legal issues* (6th ed.). Portland, Oregon: Stoel Rives LLP.

Strasburg, S. (2012). North Hill gets housing relief. *Minot Area Development Council's Newsletter.* Retrieved March 1, 2013, from http://www.minotusa.com/news

Stringham, R., & Lee, C. (2011). *Co-op survival rates in Alberta.* Retrieved March 1, 2013, from http://auspace.athabascau.ca/bitstream/2149/3132/1/BALTA%20A11%20Report%20-%20Alberta%20Co-op%20Survival.pdf

Taylor, K. (2015). Learning from the co-operative institutional model: How to enhance organizational robustness of third sector organizations with more pluralistic forms of governance. *Administrative Sciences, 5*(3), 148.

Tellis, W. (1997). Introduction to case study. *The Qualitative Report, 3*(2), 1–11.

Thornton, P. H. (2002). The rise of the corporation in a craft industry: Conflict and conformity in institutional logics. *Academy of Management Journal, 45*(1), 81–101. doi:10.2307/3069286

Thornton, P. H., & Ocasio, W. (1999). Institutional logics and the historical contingency of power in organizations: Executive succession in the higher education publishing industry, 1958–1990. *American Journal of Sociology, 105,* 801–844. doi:10.1086/210361

Thornton, P. H., Ocasio, W., & Lounsbury, M. (2013). *The institutional logics perspective: A new approach to culture, structure and process.* Oxford, England: Oxford University Press.

Tiebout, C. (1956). A pure theory of local expenditures. *Journal of Political Economy, 64*(5): 416–424, doi:10.1086/257839

Tocqueville, A. de (2006). *Democracy in America.* New York, New York: HarperCollins.

Tolbert, C. M., Irwin, M. D., Lyson, T. A., & Nucci, A. R. (2002). Civic community in small-town America: How civic welfare is influenced by local capitalism and civic engagement. *Rural Sociology, 67,* 90–113. doi:10.1111/j.1549-0831.2002.tb00095.x

Tracy, R. (2012). Renewable firms seek tax-equity partners. WSJ.com. Retrieved March 1, 2013, from https://www.wsj.com/articles/SB10001424052970203646004577215163265069078

University of Illinois Extension. (2013). McLean County agriculture history and facts [Fact sheet]. Retrieved March 1, 2013, from http://web.extension.illinois.edu/lmw/cat82_1897.html

University of Wisconsin Center for Cooperatives. (2007) Rural electric | Research on the economic impact of cooperatives [Fact sheet]. Retrieved August 5, 2011, from http://reic.uwcc.wisc.edu/electric?/

U.S. Census Bureau. (2015). QuickFacts from the U.S. Census Bureau. Retrieved August 28, 2015, from http://quickfacts.census.gov/qfd/states/17/17113.html

U.S. Council of Economic Advisers. (2016). Strengthening the rural economy—The current state of rural America. Retrieved September 1, 2016, from https://www.whitehouse.gov/node/11491

USDA Rural Development. (2000). *USDA program brought electricity and a better way of life to rural America—When the lights came on.* Retrieved August 5, 2011, from http://www.rurdev.usda.gov/rbs/pub/aug00/light.htm

USDA Rural Development. (2012). USDA and Verendrye Electric help build childcare center [Press release]. Retrieved March 1, 2013, from http://www.rurdev.usda.gov/STELPRD4016685.html

U.S. Department of Labor, Bureau of Labor Statistics. (2015). Bloomington-Normal, IL economy at a glance. Retrieved August 28, 2015, from http://www.bls.gov/eag/eag.il_bloomington_msa.htm

U.S. Energy Information Administration. (2016). How much U.S. electricity is generated from renewable energy? *Energy in Brief.* Retrieved September 1, 2016, from https://www.eia.gov/energy_in_brief/article/renewable_electricity.cfm

U.S. Environmental Protection Agency. (2009). Renewable portfolio standards fact sheet [Fact sheet]. Retrieved August 5, 2013, from http://www.epa.gov/chp/state-policy/renewable_fs.html

Veblen, T. (1994). *Absentee ownership: Business enterprise in recent times: The case of America* (Vol. 9). New Brunswick, New Jersey: Transaction Publishers.

Verendrye Electric Cooperative. (2013). About us: Board of directors [Fact sheet]. Retrieved March 1, 2013, from http://www.verendrye.com/about/board-of-directors/

Verendrye Network News. (2009). Recognize any names? Call VEC if you do. *Verendrye Network News, 8,* 14–16.

Verendrye Network News. (2011). Verendrye to waive charges for accounts without power. Retrieved March 1, 2013, from www.verendrye.com/uploads%5Cresources%5C75%5Cpage-5.pdf

Vilsack, T., & Chu, S. (2010). The rural advantage in energy independence [Blog post]. Retrieved August 5, 2011, from http://www.dailyyonder.com/when-secretaries-speak-rurals-role-energy/2010/05/28/2769

Wilkinson, K. P. (1972). A field-theory perspective for community development research. *Rural Sociology, 37*, 43–52.

Wilkinson, K. P. (1991). *The community in rural America*. New York, New York: Praeger.

Wilkinson, R., & Pickett, K. E. (2009). *The spirit level: Why more equal societies almost always do better*. Harmondsworth, England: Allen Land.

Windustry. (2013). Passive tax appetite [Fact sheet]. Retrieved March 1, 2013, from http://www.windustry.org/passive-tax-appetite

World Wind Energy Association. (2010). *World wind energy report 2009*. Retrieved August 5, 2011, from www.wwindea.org/home/images/.../worldwindenergyreport2009_s.pdf

World Wind Energy Association. (2012). *2012 half-year report*. Retrieved March 1, 2013, from www.wwindea.org/webimages/Half-year_report_2012.pdf

World Wind Energy Association. (2015). Worldwide wind market booming like never before: Wind capacity over 392 gigawatt. Retrieved September 1, 2016, from http://www.wwindea.org/hyr2015/

Yin, R. K. (2009). *Case study research: Design and methods* (4th ed., Vol. 5). Thousand Oaks, California: Sage.

Zacharakis, J., & Flora, J. (2005). Riverside: A case study of social capital and cultural reproduction and their relationship to leadership development. *Adult Education Quarterly, 55*, 288–307. Doi:10.1177/0741713605277370

Zeuli, K., & Radel, J. (2005). Cooperatives as a community development strategy: Linking theory and practice. *Journal of Regional Analysis and Policy, 35*(1), 43–54.

Zimbelman, C. (2013). Year-end editorial from the mayor of Minot [Blog post]. Retrieved March 1, 2013, from http://www.minotrecoveryinfo.com/news/detail.asp?newsID=252

INDEX

CPSIA information can be obtained
at www.ICGtesting.com
Printed in the USA
FSHW020322300719
60519FS